Disclaimer

The publisher of this book is by no way associated with the National Institute of Standards and Technology (NIST). The NIST did not publish this book. It was published by 50 page publications under the public domain license.

50 Page Publications.

Book Title: Comparison of the WSQ and JPEG 2000 Image Compression Algorithms On 500 ppi Fingerprint Imagery

Book Author: John M. Libert; Shahram Orandi; John D. Grantham

Book Abstract: This paper presents the findings of a study conducted to compare the effects of WSQ and JPEG 2000 compression on 500 ppi fingerprint imagery at a typical operational compression rate of 0.55 bpp (bits per pixel), corresponding to an effective compression ratio of approximately 15:1. Compression effects are measured using peak signal to noise ratio (PSNR), proportion of pixels changed via compression regardless of magnitude, and a frequency analytic method the Spectral Image Validation/Verification (SIVV) metric. Additional information regarding compression effects is provided by comparison of fingerprint minutia templates extracted for uncompressed and compressed versions of fingerprints. The study also examines comparative behavior of the two compression CODECs with respect to multiple compression cycles, recompression of images using dissimilar CODECs, and compression fidelity error as a function of compression bit rate.

Citation: NIST Interagency/Internal Report (NISTIR) - 7781

Keyword: Fingerprint compression; wavelet scalar quantization; WSQ; JPEG 2000 fingerprint compression; Fingerprint compression; wavelet scalar quantization; WSQ; JPEG 2000 fingerprint compression; 500 ppi fingerprint imagery; SIVV; PSNR

Comparison of the WSQ and JPEG 2000 Image Compression Algorithms On 500 ppi Fingerprint Imagery

NIST Interagency Report 7781

John M. Libert, Shahram Orandi, John D. Grantham

Image Group

Information Access Division

Information Technology Laboratory

National Institute of Standards and Technology

April 24, 2012

ACKNOWLEDGEMENTS

The authors wish to give special thanks the following individuals and organizations for their support of this work:

    Federal Bureau of Investigation for all their support throughout this study
    Margaret Lepley of MITRE Corporation

In addition, we appreciate the guidance, support and coordination provided by Michael Garris without whose help and support this study would not have been possible.

DISCLAIMER

Specific hardware and software products identified in this report were used in order to perform the evaluations described in this document. In no case does identification of any commercial product, trade name, or vendor, imply recommendation or endorsement by the National Institute of Standards and Technology, nor does it imply that the products and equipment identified are necessarily the best available for the purpose.

Executive Summary

The criminal justice communities throughout the world exchange fingerprint imagery data primarily in 8-bit grayscale and at a resolution of 500 pixels per inch [1] (ppi) or 19.7 pixels per millimeter (ppmm). The Wavelet Scalar Quantization (WSQ) fingerprint image compression algorithm is currently the standard algorithm for the compression of 500 ppi fingerprint imagery in the United States. WSQ is a "lossy" compression algorithm. Lossy compression algorithms employ data encoding methods which discard (lose) some of the data in the encoding process in order to achieve a reduction in the digital file size of the data being compressed. Decompressing the resulting compressed data yields output that, while different from the original, is similar enough to the original that it remains useful for the intended purpose. The WSQ algorithm allows for users of the algorithm to specify how much compression is to be applied to the fingerprint image at the cost of increasingly greater loss in fingerprint image fidelity as the effective compression ratio is increased (see Figure 1 for an example of image degradation from lossy compression). The WSQ Grayscale Fingerprint Image Compression Specification [WSQ] provides guidance based on an International Association for Identification (IAI) study [FITZPATRICK] to determine the acceptable amount of fidelity loss due to compression in order for a WSQ encoder and decoder to meet FBI certifications. These certifications are designed to ensure adherence to the WSQ specification and thereby to ensure fidelity and admissibility in courts of law for images that have been processed by such encoders and decoders.

For 1000 ppi (39.4 ppmm) fingerprint imagery, MITRE has developed an informative guidance ("Profile for 1000 ppi Fingerprint Compression" [MTR1]) that is widely recognized as the *de facto* standard guidance for utilizing JPEG 2000 [2] for the compression of fingerprint imagery at 1000 ppi. This document provides a basis for a compression profile for 1000 ppi fingerprint imagery using JPEG 2000, particularly in its specification of software parameters for control and structure of the JPEG 2000 code stream, and its recommendations form the basis for the compression strategy used in this study. Note that JPEG 2000 is also a lossy compression algorithm.

Preliminary examination of image spectra indicated that while JPEG 2000 serves as a worthy successor to WSQ in transition from 500 ppi to 1000 ppi, there may be behavioral differences between the two algorithms that need to be studied further and that these behavioral differences may be more pronounced in cases where the combination of both algorithms are used such as when images are rescaled from a higher resolution (1000 ppi, compressed with the *de facto* compression standard for this type of imagery, JPEG 2000) to a lower resolution (500 ppi and recompressed with WSQ). The experimental design for the present study took into account typical operational scenarios where WSQ is utilized for 500 ppi and JPEG 2000 is utilized for 1000 ppi. This study also attempted to look at the intrinsic differences between the two algorithms on equal footing without image resolution biasing the results. Since WSQ is designed to only operate at 500 ppi and JPEG 2000 offers flexibility to be operated at both 500 ppi and 1000 ppi, comparison of algorithm behavior on equal footing can take place at 500 ppi.

The study shows that the WSQ and JPEG 2000 CODECs exhibit fundamental differences in behavior. The key differences appear to be linked to how the algorithms operate within various image frequency ranges. For example, WSQ focuses on retaining lower frequency components (i.e., fingerprint ridge structure and minutiae) more effectively at the cost of degrading more rapidly at the high frequency features (i.e., paper fibers). JPEG 2000 on the other hand exhibits a more linear degradation across the frequency bands with less selective loss at the higher frequency range. The frequency behaviors of the respective algorithms seemed to be inversely related to variability measures such as peak signal to noise ratio (PSNR), or the count of pixels changed pre- and post-processing, using the algorithms in question. This finding suggests that using PSNR or calculating the number of changed pixels alone may not be the best method to use for measuring or comparing degradation from compression.

---

[1] Resolution values for fingerprint imagery are specified in pixels per inch (ppi) throughout this document. This is based on widely used specification guidelines for such imagery and is accepted as common nomenclature within the industry. SI units for these will be presented only once.

[2] The "2000" refers to the year of publication of the first edition of the image compression standard known as JPEG 2000. JPEG refers to Joint Photographic Experts Group.

In cases more closely mimicking operational scenarios, such as an incoming 1000 ppi JPEG 2000 image arriving at the processing center, being decompressed, down-sampled to 500 ppi and finally recompressed using WSQ, the system appeared to fare better than the hypothetical reverse of this case; where the image starts at WSQ, is decompressed, and recompressed with JPEG 2000. It should be noted however that depending on the metric, the penalty incurred by using two different algorithms (i.e., WSQ, then JPEG 2000) is larger than that of two passes of the same algorithm, so the combination of both algorithms does incur a larger cumulative effect.

With regard to progressive degradation through multiple passes of compression using the same algorithm, it was shown that JPEG 2000 offers the most stability where the degradation loss is incurred in the first pass of compression and then stabilizes with no further statistically significant measured loss. WSQ on the other hand demonstrates that a small but statistically significant loss occurs on the second pass of compression, and that fidelity continues to diminish over successive compression cycles before PSNR reaches some minimum level.

VERSION HISTORY

Date	Activity
01/09/2012	Draft v.1

TABLE OF CONTENTS

1. Investigative Goals and Objectives 19
   - 1.1. Background 19
   - 1.2. WSQ to JPEG 2000 Comparison 21
2. Materials and Methods 23
   - 2.1. Image Data 23
   - 2.2. Compression Algorithms 23
   - 2.3. Methodology 25
3. Analysis 29
   - 3.1. Normality of Metrics 29
   - 3.2. Uncertainty of Median 29
   - 3.3. Hypothesis Testing 29
4. Results 31
   - 4.1. Investigative Goal 1: Compare WSQ to JPEG 2000 in terms of fidelity to the original image 31
   - 4.2. Investigative Goal 2: Examine effects of multiple compression cycles 39
5. Conclusions 43
   - 5.1. Fidelity: Single Compression Cycle 43
   - 5.2. Fidelity: Two compression cycles each using either WSQ or JPEG 2000 CODECs 43
   - 5.3. Fidelity: Two compression cycles using dissimilar encoders 43
   - 5.4. Effects of compression/decompression over extreme numbers of multiple cycles 43
6. Future Work 45

References 47
   - Publications and Reports 47
   - Standards 49

Appendix A. Compression Ratio 51
Appendix B. Effect of Image Contrast on SIVV Power Signal 53
Appendix C. Distributions of Comparison Measurements 55

LIST OF TABLES

Table 1 - Abbreviations 15
Table 2 - JPEG 2000 Compressor Settings Used in Study 24
Table 3 - Algorithm Combinations Tested 31
Table 4 – Median values of fidelity metrics for each of 7 compression treatment cases, each having N=2898 values of each of the three metrics. 32
Table 5 – Simultaneous Hypothesis Tests of Differences Between Processing Treatments ($\alpha=0.05$) 35
Table 6 - Median compression ratios and % normalized differences with upper and lower 95 % confidence limits ($N=2898$) 52

List of Figures

Figure 1 - Example of Fidelity Degradation Due to Extreme Lossy Compression (JPEG 2000 at 800:1) 20
Figure 2 - Flow of compression processing with numerals labeling image outputs of encode/decode cycles. 25
Figure 3 - NIST SIVV metric applied to non-compressed original image and to decoded WSQ compressed image. 28
Figure 4 - Median values of three metrics with individual 95 % confidence intervals on the statistic for the seven WSQ and JPEG 2000 compression conditions examined in the present investigation for each of the three metrics. 32
Figure 5 - Example of typical behavioral differences between JPEG 2000 and WSQ (15:1 compression) 33
Figure 6 - Median values of three metrics with 95 % confidence intervals on the statistic for seven WSQ and JPEG 2000 compression conditions examined in the present investigation. (Note that larger power differences indicate lower fidelity to non-compressed original image). 36
Figure 7 - Median PSNR and 95 % confidence intervals for 1000 ppi, inked fingerprint images classified by expert examiners into each of the degradation categories 1 – 3. Only matching pairs of non-compressed and compressed versions of the same image are included here; i.e. same subject, same finger. 37
Figure 8 - Median power difference by frequency band with individual 95 % confidence intervals for 1000 ppi fingerprints subjected to various levels of JPEG 2000 compression grouped by ratings assigned by expert fingerprint examiners relative to degradation of features useful for identification. Total power difference is shown as well as sum of difference over each of 5 frequency bands from 0.0 cycles/pixel to 0.5 cycles/pixel. 38
Figure 9 - Median Peak Signal to Noise Ratio over 200 compression cycles applied to each of 200 fingerprint images. In contrast with WSQ, JPEG2000 exhibits remarkable stability even at such extremes of repeat compression/decompression. (In each case, PSNR compares the multiple compression cycle output to the non-compressed original).. 39
Figure 10 – Images at upper left and right show respectively one of the non-compressed original images used in the experiment and the result of 200 cycles of compression/decompression with the WSQ CODEC. The lower graph shows the spectra of the non-compressed image, the result of a single cycle of WSQ compression, and that of 200 cycles. PSNR values (in decibel units) appear in the legend. The spectral pattern is similar for 1 cycle and 200 cycles but power at all frequencies is reduced. 40
Figure 11 – Images at upper left and right are respectively one of the non-compressed original images and that resulting from 200 cycles of compression/decompression using the JPEG 2000 CODEC. The lower plot shows that while spectral power degradation is evident as early as in the middle of the frequency range and increases toward the high frequency, this degradation is relatively stable even up to 200 cycles of compression and is virtually identical to that after only a single cycle of compression using this CODEC. PSNR difference is negligible as well. 41
Figure 12 – Non-compressed image (upper left) has had its contrast enhanced using an adaptive histogram equalization algorithm (upper right). The lower display of the SIVV signals shows the almost uniform increase in power over all frequencies. Note the low PSNR value as a result of this operation in spite of the high correlation of the frequency structure of the two images. 54
Figure 13 - Boxplots of distributions of PSNR for each compression experiment. 55
Figure 14 - Boxplots of distributions of proportion of altered pixels for each compression experiment. 55
Figure 15 - Boxplots of distributions of total difference in power spectrum for each compression experiment. 56

TERMS AND DEFINITIONS

The abbreviations and acronyms of Table 1 are used in many parts of this document.

Table 1 - Abbreviations

CODEC	Coder/decoder (or compression/decompression algorithm, module, software)
FBI	Federal Bureau of Investigation
IAFIS	Integrated Automated Fingerprint Identification System
IAI	International Association for Identification
JP2K	JPEG 2000
JPEG	Joint Photographic Experts Group – ISO/IEC committee developing standards for image compression – also used as the name of the CODEC developed in accordance with the standard specified by this body.
NBIS	NIST Biometric Image Software
NGI	Next Generation Identification
NIST	National Institute of Standards and Technology
SIVV	Spectral Image Validation/Verification metric
WSQ	The Wavelet Scalar Quantization algorithm for compression of fingerprint imagery
ppi	Pixels per inch
bpp	Bits per pixel
PSNR	Peak Signal To Noise Ratio
ppmm	Pixels per millimeter
DCT	Discrete Cosine Transform
OPJ	OpenJPEG's JPEG 2000 CODEC
CR	Compression Ratio

ABSTRACT

This paper presents the findings of a study conducted to compare the effects of WSQ and JPEG 2000 compression on 500 ppi fingerprint imagery at a typical operational compression rate of 0.55 bpp (bits per pixel), corresponding to an effective compression ratio of approximately 15:1. Compression effects are measured using peak signal to noise ratio (PSNR), the proportion of pixels changed via compression regardless of magnitude, and the Spectral Image Validation/Verification (SIVV) metric, a frequency analytic method. The study also examines comparative behavior of the two compression algorithms with respect to multiple compression cycles, recompression of images using dissimilar CODECs, and compression fidelity error as a function of compression bit rate. The study finds that while JPEG 2000 exhibits a small advantage over WSQ with respect to PSNR comparison, frequency spectrum comparison shows that WSQ is better tuned to the preservation of fingerprint features than JPEG 2000. However, JPEG 2000 is considerably more stable over multiple compression cycles.

KEYWORDS

Fingerprint compression; wavelet scalar quantization; WSQ; JPEG 2000 fingerprint compression; 500 ppi fingerprint imagery; SIVV; PSNR

1. Investigative Goals and Objectives

In July of 2009 National Institute of Standards and Technology (NIST) in partnership with the Federal Bureau of Investigation (FBI) commenced an investigation on the use of JPEG 2000 [JPEG2K] for compressing fingerprint imagery with the following objectives:

1. **Compare WSQ to JPEG 2000 in terms of fidelity to the non-compressed original image:**
    a) **Single compression cycle using either WSQ or JPEG 2000 CODEC:** Compressing 500 ppi fingerprints at a target rate of 15:1 using each of the CODECs, compare peak signal to noise ratio (PSNR) as well as changes in spectral power with respect to the power spectrum of the original image.
    b) **Two compression cycles each using either WSQ or JPEG 2000 CODECs:** Measure progressive effects on fingerprint images of multiple compression/decompression cycles with WSQ and with JPEG 2000 CODEC.
    c) **Two compression cycles using dissimilar encoders:** Measure effects of compressing and decompressing using one encoder and recompressing using the other.
2. **Examine the effects of compression/decompression over multiple cycles:** While not likely to be practiced, compression, decompression, and recompression of images over a high numbers of cycles can reveal useful information concerning practical limitations and stability of the CODEC. Hence, we examine change in PSNR with successive recompression with WSQ and JPEG 2000 CODECs.

1.1. Background

The criminal justice communities throughout the world exchange fingerprint imagery data primarily in 8-bit grayscale at a resolution of 500 pixels per inch (ppi). The Wavelet Scalar Quantization (WSQ) [BRADLEY1], [BRADLEY2], [BRISLAWN], [HOPPER] fingerprint image compression algorithm is currently the standard algorithm for the compression of 500 ppi fingerprint imagery in the United States. The WSQ standard defines a class of encoders and decoders with sufficient interoperability to ensure that images encoded by any compliant encoder can be decoded by any other compliant decoder.

Since the WSQ algorithm was specifically designed for 500 ppi fingerprint imagery, it was neither validated nor prescribed for use at resolutions beyond 500 ppi fingerprint imagery. With the introduction of 1000 ppi capture of fingerprints as well as other friction ridge biometrics, e.g. palms, the WSQ algorithm has been replaced with the JPEG 2000 algorithm which can operate effectively across a variety of image formats, resolutions and color depths. Hence, while not designed specifically for (or tuned specifically for) the compression of fingerprints, the modern JPEG 2000 [JPEG2K] algorithm provides all the necessary flexibility to allow its utilization in the compression of friction ridge imagery. It should also be noted that like WSQ, JPEG 2000 is also a wavelet based image transformation algorithm, but unlike WSQ, JPEG 2000 implementations are more widely available as general purpose commercial off-the-shelf (COTS) products as well as several no-cost public domain implementations. For high fidelity compression of still images, JPEG 2000 is rapidly making inroads in replacing its predecessor, the discrete cosine transform (DCT) based JPEG [JPEG] algorithm (also referred to as baseline JPEG).

Both WSQ and JPEG 2000 are "lossy" [3] compression algorithms. Lossy compression algorithms employ data encoding methods that discard (lose) some of the data in the encoding process in order to achieve a reduction in the storage space required by the data being compressed. Decompressing the resulting compressed data yields output that, while different from the original, is similar enough to the original that it remains useful for the intended purpose. *Lossless* compression algorithms on the other hand can produce a compressed image that can be decompressed back to original form with no loss or change to the resulting image. The disadvantage to lossless algorithms is that they produce compressed images that can be many times larger in file size than compressed images produced by lossy algorithms.

---

[3] JPEG 2000 has a lossless mode able to achieve a compression ratio of approximately 1.5:1, but the main interest in this CODEC for biometric samples involves only the lossy mode.

The lossy compression algorithms allow for users of the algorithms to specify how much compression to apply to the fingerprint image at the cost of increasingly greater loss in fingerprint image fidelity as the effective compression ratio is increased (See Figure 1 as an example of image degradation from extreme lossy compression). The WSQ Grayscale Fingerprint Image Compression Specification [WSQ] provides guidance for the acceptable amount of fidelity loss due to compression in order for the encoder and decoder to meet FBI certifications for 500 ppi fingerprint imagery. These certifications are designed to ensure adherence to the WSQ specification to ensure sufficient fidelity for admissibility in courts of law for images that have been processed by such encoders and decoders.

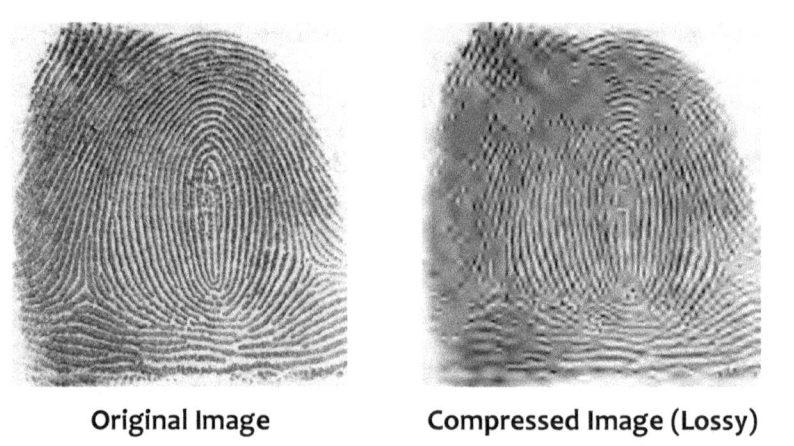

Original Image **Compressed Image (Lossy)**

Figure 1 - Example of Fidelity Degradation Due to Extreme Lossy Compression (JPEG 2000 at 800:1)

A study conducted by the International Association for Identification (IAI) [FITZPATRICK] established 15:1 as a WSQ compression ratio that would retain acceptable image fidelity in 500 ppi fingerprint imagery. The study used the judgments of expert fingerprint examiners to measure the fidelity loss due to compression. In order to reduce bias due to subjectivity, multiple examiner decisions were used to build consensus. Utilizing examiners' opinions does not imply that automated fingerprint matcher performance is not an important criterion in a given biometric system, but it must be noted that if fingerprints are to be admissible as evidence in a court of law their ultimate utility lies in the expert examiner's opinion of the fidelity of those fingerprints.

Current NIST research, including the study being presented here, is directed toward the goal of establishing a normative guidance (such as that described for WSQ by the IAI) for JPEG 2000. In establishing the basis for a normative guidance, NIST has conducted several studies examining various behavioral aspects of the JPEG 2000 algorithm relative to human evaluation of feature degradation [NISTIR7778] or machine determination of identification probability.

Whereas the other studies [IAI] and [NISTIR7778] examined the various factors impacting the behavior of WSQ and JPEG 2000 respectively under various scenarios with respect to visual responses of trained fingerprint examiners, the present study aims to explore quantitative differences between WSQ and JPEG 2000 algorithms with respect to signal fidelity. Most of this study focuses on scenarios typical of an operational system, but also expands its scope to some regimes which would normally not be encountered. While approaching some of these boundary conditions in normal operations may be unlikely, the extreme conditions presented to the algorithms in this study are important in testing the limits of CODEC functionality and can reveal behavioral characteristics that would otherwise not be easily visible under less rigorous circumstances. The experiments are conducted to broaden the base of information, both supportive and cautionary, relative to the decision to adopt JPEG 2000 as the new standard for compression of friction ridge biometric data.

1.2. WSQ to JPEG 2000 Comparison

Lepley [MTR1] compares JPEG 2000 to WSQ fingerprint compression mainly from the point of view of interoperability. The Lepley study mainly examines potential compatibility of the code streams and possibilities for transcoding which usually refers to recoding the compressed image data to an alternate format without passing through a complete decompression and recompression cycle. Lepley elaborates on algorithmic differences between the two CODECs that account for difficulties in transcoding. Lepley observes differences in fingerprint image quality between the algorithms and suggests recommendations for modifications to the JPEG 2000 standard to better replicate WSQ behavior and to enhance potential for compatibility at the code stream level. Comparing images compressed with the two CODECs, she finds that JPEG 2000 images show higher peak signal to noise ratio [KUMAR] when compared to non-compressed original images, but subjective evaluation of processed images suggests higher quality for WSQ processed images. JPEG 2000 images are reported to exhibit a softer appearance in comparison to the crisper appearance and better fine texture rendition of WSQ. Retention of ridge, bifurcation, and sweat pore details are found to be similar for both algorithms. In addition to comparison of PSNR and visual inspection/subjective evaluation, Lepley employs a quality measurement tool, Image Quality Metric IQM [NILL4] that evaluates digital image quality with respect to the weighting of spatial frequency characteristics of the image. This metric indicates better performance by WSQ, and may support the visually observed loss of fine image detail through subjective evaluation.

Figueroa-Villanueva, et al. [FIG-VILL] measure PSNR and fingerprint verification match scores of images from three publicly available fingerprint databases compressed using WSQ and JPEG 2000 CODECs. They observe that given the same target compression rate, the effective compression rates of WSQ output images can vary significantly, whereas JPEG 2000 compression rates were very close to the target rates set in the CODEC. They attempt to correct for disparities in effective compression rates by calculating the average effective compression rate of the images output by the WSQ CODEC and using that average effective rate value as the target compression rate parameter for the JPEG 2000 CODEC. Using target compression rates of 2.25 bpp and 0.75 bpp they observe lower distortion (and higher PSNR) for JPEG 2000 than for WSQ. Moreover, they find that genuine match scores (in same finger, one to one matches) to be higher for JPEG 2000 than for WSQ. Both tests are reported by the authors to be statistically significant.

Funk, et al. [FUNK] compare the performance of three automated fingerprint matchers on images compressed using JPEG, WSQ, and JPEG 2000 CODECs. They apply compression to each image in a suite of 2170 fingerprint images acquired from 30 subjects over 3 days, and gradually increase the compression rate from near lossless levels to around 38:1 ($\approx$0.21 bpp). They apply three automated matchers to determine the effect of lossy compression on fingerprint images. They report very close agreement among the three CODECs for match scores up to a compression rate of around 18:1 (0.44 bpp), with mean match scores (average match probabilities) ranging [0.993, 0.997]. Beyond 18:1 effective compression, JPEG scores drop precipitously whereas close agreement between WSQ and JPEG 2000 persists throughout the remaining range of compression ratios.

2. Materials and Methods

2.1. Image Data

A newly scanned version, SD27A, of the NIST Special Database 27 [SD27] is used for the present study. The "ten-print" portion of this database consists of 2898 fingerprint images scanned from standard FBI ten-print inked fingerprint cards at scan rates of 500 ppi, 1000 ppi, and 2000 ppi. Scans at each respective resolution were made without changing the fingerprint card's position on the scanner platen. Thus, three images of the same fingerprint are captured, at three sample rates (500, 1000 and 2000 ppi). The standard ten-print card includes ink-rolled fingerprints of each of the ten fingers, four- finger flat (slap) impressions of each hand, and flat impressions of the left and right thumbs for a total of 14 impressions per card for each of 207 individual fingerprint records.

2.2. Compression Algorithms

The WSQ (Wavelet Scalar Quantization) algorithm is currently the fingerprint image compression standard prescribed by the FBI and used for both storage and transmission of 500 ppi 8-bit grayscale fingerprint imagery. The algorithm was developed specifically for fingerprints as part of the effort to convert the FBI's fingerprint database to digital format. Early tests [MIL] found the discrete cosine transform compression scheme specified by the Joint Photographic Experts Group [JPEG] to be not as well suited as WSQ for fingerprint compression due to the presence of blocking artifacts and the degradation of minutiae. At the time of WSQ development by the FBI, other compression schemes were considered and tested as well. WSQ was originally developed as the "simplest possible wavelet algorithm" and intended as a comparison baseline for more complex algorithms [HOPPER]. The computational details of WSQ are provided in [BRADLEY1], [BRADLEY2], and [BRISLAWN].

JPEG 2000 [JPEG2K], [TAUBMAN1], and [TAUBMAN2] is an image compression standard and coding system that was created to improve upon the original JPEG image compression standard's discrete cosine transform-based methodology [JPEG] by utilizing a wavelet-based encoding methodology. This modification yielded increases in both effective data compression rates and subjective image quality. Moreover, JPEG 2000 provides additional flexibility in the creation and manipulation of the compressed data code-stream and is based on the same family of wavelets as WSQ which is currently the standard for fingerprint image compression at 500 ppi. The flexibility offered by JPEG 2000 as well as the greater availability of JPEG 2000 implementations, which are commodity products as opposed to the much more specialized WSQ implementations, make JPEG 2000 a good candidate for succession from WSQ in emerging high-resolution/large-geometry use-cases such as palm imagery or 1000 ppi fingerprint images.

The implementation of JPEG 2000 used in this experiment is OpenJPEG's [OPENJPEG] reference implementation version 1.4. This reference implementation has been incorporated into the NIST Biometric Image Software (NBIS) public domain software distribution [NIST2] release 4.0.0. The implementation of WSQ used in this experiment is the reference implementation developed by NIST and available in the NBIS release 4.0.0.

Note that JPEG 2000 enables structuring of the code stream such that it may be progressively decompressed at any of a series of intermediate compression levels in addition to the final "target" compression level. This feature of JPEG 2000 is intended to allow for the display of lower fidelity versions of the image to suit, for example, lower resolution displays while adding negligibly to the size of the compressed data stream and having no effect on the image at the target compression level. This feature was not used in the present experiments, hence only a single target compression rate was specified to the encoder. Other specific configuration parameters used for the compression of images are provided in Table 2.

Table 2 - JPEG 2000 Compressor Settings Used in Study

Compressor Configuration Setting	Description
-n 6	6 resolution levels (original + 5 levels of decomposition[4])
-p RPCL	Resolution-Position-Component-Layer (RPCL) progression order
-b 64, 64	Code block size of 64x64
-r []	Specifies the target top-layer rate, plus other quality layers
-d 0,0	Image origin offset
-I	Use irreversible compression (lossy)
-S 1,1	Use subsampling factor of 1,1

---

[4] The OpenJPEG 2000 CODEC sets the number of decomposition levels to one less than the value specified by this command line parameter. Hence, -n 6 yields 5 decomposition levels.

2.3. Methodology

2.3.1. Compression Processing: Operational Scenarios

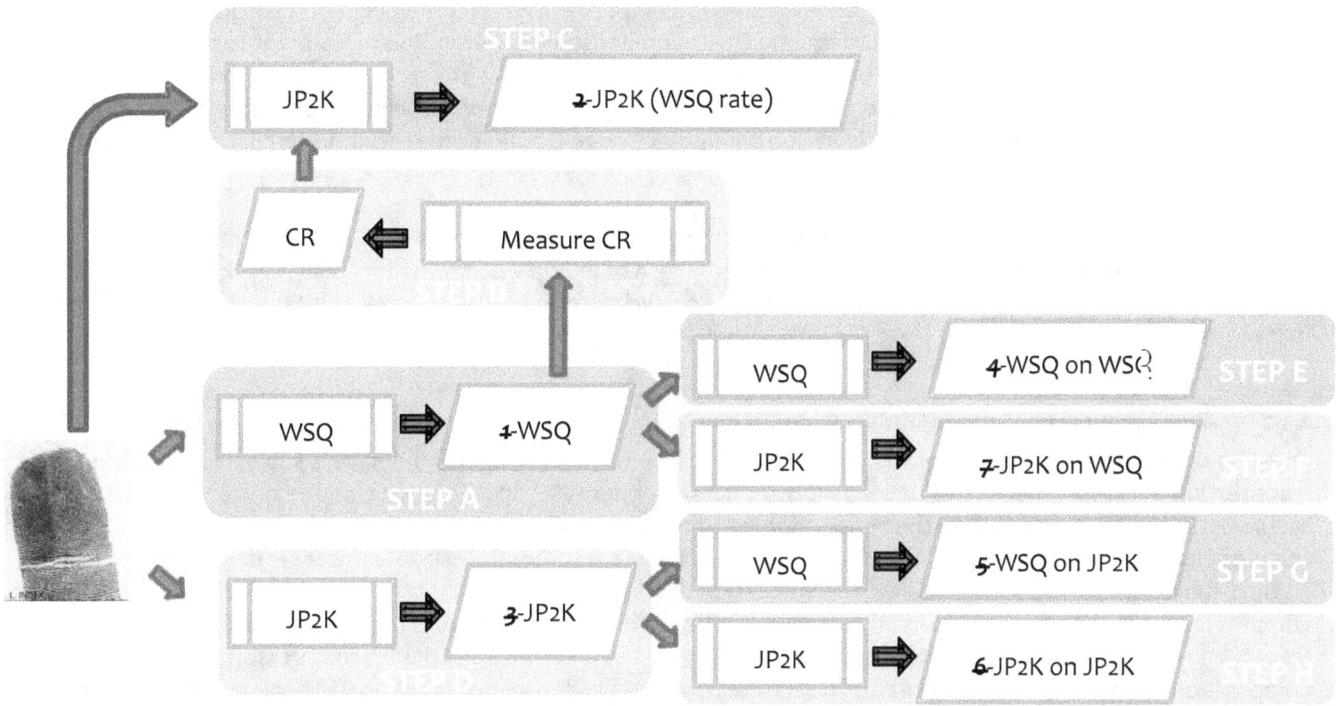

Figure 2 - Flow of compression processing with numerals labeling image outputs of encode/decode cycles.

Figure 2 depicts the stages of processing using the WSQ and JPEG 2000 CODECs. Each original fingerprint image is subjected to the following steps[5]:

A. Encode and decode each image using an FBI/NIST-certified version of WSQ [NBIS] using a target bit rate of 0.55[6] bpp to yield image **1** WSQ.

B. Measure the effective compression ratio of the compressed WSQ file compared to the size of the file input to the WSQ CODEC to yield a Compression Ratio (CR).

C. Encode and decode each image using JPEG 2000 CODEC from OpenJPEG 2000 version 1.4 [JPEG2K] specifying a target compression ratio equal to that measured for the WSQ image, i.e., CR, to yield image **2** JP2K (WSQ rate).

D. Encode and decode each image using JPEG 2000 encoder/decoder from OpenJPEG 2000 version 1.4 specifying a target compression ratio of 15:1 to yield the output image **3** JP2K.

E. Apply the WSQ CODEC to image **1** WSQ to yield image designated **4** WSQ on WSQ.

F. Apply the OpenJPEG 2000 CODEC to image **1** WSQ to yield image designated **7** JP2K on WSQ.

G. Apply the WSQ CODEC to image **3** JP2K to yield image designated **5** WSQ on JP2K.

H. Apply the OpenJPEG 2000 CODEC to image **3** JP2K to yield the output image designated as **6** JP2K on JP2K.

---

[5] Note: Processing steps are designated with letters and outputs are specified with numbers.
[6] Note that in order to achieve the average compression ratio of approximately 15:1, the FBI typically uses a bit rate specification of 0.75 [BRISLAWN2] rather than the 0.55 bpp used here. We use the published compression ratio of approximately 15:1 (i.e. 0.55 bpp) in order to examine variability of WSQ given this intended compression ratio as a target (see Appendix A.)

Note that the actual compression rate achieved by the WSQ CODEC can vary considerably about an intended target. Thus, specifying a bit rate such as 0.55 bpp (≈14.5:1) might generate actual compression ratios that vary considerably (see Appendix A). Given a compression ratio target, such as 15:1, JPEG 2000 CODECs, by contrast, generate compressed outputs very close to the specified target rate. To provide a controlled frame of reference in the observation of compression degradation, JPEG 2000 was applied to each fingerprint using both a general/fixed target compression rate target as well as a floating target compression ratio obtained by measuring the resulting compression rate from the WSQ output to facilitate direct comparison with WSQ. This adjustment to the compression target ratio for JPEG 2000 is not realistic operationally, but was done to allow the most uniform and complete comparison between CODECs known to differ with respect to observed compression rate given the same input data by effectively locking the target rate configuration parameter of one CODEC to the resulting compression rate of the other CODEC.

The procedure described above includes cases **4** and **6**, representing respectively two cycles of compression with WSQ and JPEG 2000 CODECs. To examine the extreme case of multiple compression cycles, a random sample of images were subjected to 200 compression/decompression cycles using each of the two CODECs, as described further below.

2.3.2. Comparison over Multiple Compression Cycles

The previous section describes several likely operational cases in which a fingerprint image is subjected to multiple compression/decompression cycles[7], including two cycles of either WSQ or JPEG 2000 compression and mixed CODEC scenarios. In addition, we examine the behavior of the two CODECs with extremes of multiple compression cycles. For this we select a uniform random sample of $N=200$ fingerprint images from the SD27A dataset (as described in section 2.1) and subject each image to 200 cycles of compression/decompression. Each of the sampled images is first compressed using the two CODECs, then decompressed, and the resulting image is then compressed again. The PSNR (see 2.3.3.1 below for more information) is measured between the original non-compressed image and each compression/decompression cycle output. Each of the N images yields a vector of $k=200$ PSNR values for each of the two CODECs. The experiment performed with each CODEC yields an $N \times k$ matrix of PSNR values, the columns of which represent the distribution of PSNR values for each of $k =1…200$ compression cycles. The sample median PSNR for each value k is computed and a bootstrap procedure (see 3.2) is employed to estimate the 95 % confidence interval of the median estimate.

---

[7] A "cycle" consists of compression, decompression, and then recompression of the decompressed output of the previous compression.

2.3.3. Image Fidelity Metrics

2.3.3.1. Peak Signal to Noise Ratio (PSNR)

PSNR is a commonly used measure of image fidelity [KUMAR], the degree to which one image matches another, such as in comparing a processed image to an original (non-compressed) image. Thus, given images I and K the mean squared error, MSE, between the corresponding gray level values of the two images of dimension $m \times n$ pixels is

$$MSE = \frac{1}{mn} \sum_{i=1}^{m} \sum_{j=1}^{n} [I(i,j) - K(i,j)]^2 \qquad (1)$$

and the PSNR is defined as

$$PSNR = 20 \times \log_{10} \left(\frac{Max_I}{\sqrt{MSE}} \right) \qquad (2)$$

where Max_I for 8-bit (grayscale) images is 255 (and MSE≠0)[8].

Involving only pixel-to-pixel comparison, PSNR is not always an adequate measure of image similarity, so care must be taken in its interpretation [GIROD], [HUYNH]. For example, two images differing only by some small constant may yet contain identical structural detail, yet substantially reduced PSNR. Hence, PSNR often may contradict visual assessment of image fidelity. Yet PSNR, if used cautiously, remains a useful metric of image fidelity.

2.3.3.2. Proportion of Changed Pixels

Whereas PSNR carries information regarding the magnitude of image differences, a given PSNR value may reflect either large changes in few pixels or small changes in many. Hence, a measure of the simple proportion of pixels altered by a process provides additional information as to the extent of pixel modification independent of magnitude.

---

[8] PSNR is infinite in the case that MSE=0, i.e., that the images being compared are identical.

2.3.3.3. Spectral Image Validation Verification (SIVV) Metric

Developed initially as a method to screen fingerprint databases for non-fingerprint images, segmentation errors, or mislabeled sample rates, the Spectral Image Validation Verification (SIVV) metric [LIBERT] provides a comparatively straightforward method by which to assess the frequency structure of an image. Pairwise display of the SIVV signals of non-compressed and compressed images enables summary visualization of the effects of compression across the composition frequency spectrum of the image. As a 1-dimensional representation of a 2-dimensional Fourier spectrum, the SIVV metric applied to a fingerprint image exhibits a peak corresponding to the frequency of the ridge spacing. Also, as shown in Figure 3, comparison of SIVV signals of non-compressed and compressed images shows the loss or gain of power over various frequencies.

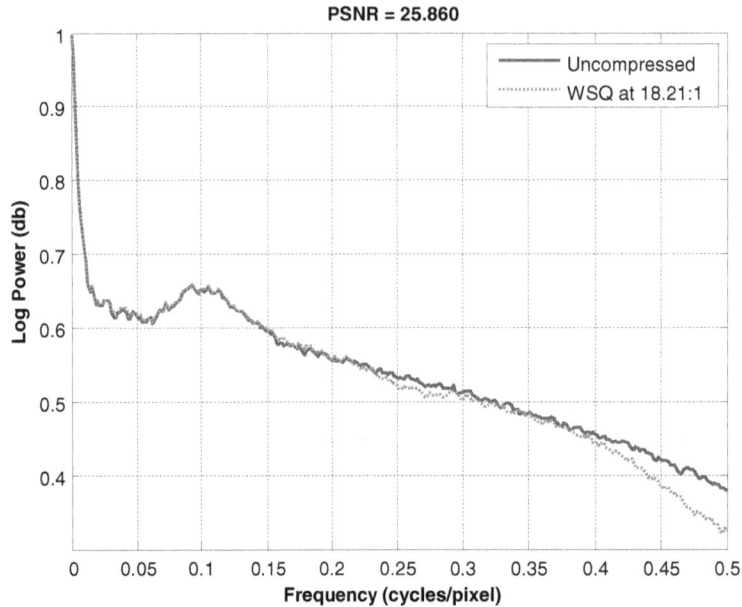

Figure 3 - NIST SIVV metric applied to non-compressed original image and to decoded WSQ compressed image.

Either differences or ratios of SIVV signals can provide quantitative measures for the comparison of compression methods. For the present study, we examine the difference in the power of non-compressed and compressed images summed over the entire signal length as well as within 5 frequency bands [9]. The power difference over the entire signal length or between corresponding frequency bands is computed as

$$dP_{f(min)}^{f(max)} = \sum_{f(min)}^{f(max)} s^{uc} - \sum_{f(min)}^{f(max)} s^{c} \qquad (3)$$

where $f(min)$ and $f(max)$ are 0.0 and 0.5 for the overall difference in signal power or the lower and upper limits of intervals of 0.1 cycles/pixel in width delineating each of the 5 frequency bands. The SIVV signals denoted as s^{uc} and s^{c} are respectively vectors of SIVV signal values for non-compressed and compressed images. The frequency samples, f, in units of cycles per pixel correspond to image pixels or Fourier transform frequencies along the length of one half of the minimum dimension of the 2D Fourier transform of the image under examination. Frequency along this dimension is scaled to the interval [0, 0.5] cycles/pixel. Note that the power value at $f=0$ is the "direct current" (DC) term, corresponding to the average intensity of the image and is used to normalize the power spectrum.

The power difference metric defined above can provide a measure of the overall difference between the compressed and original image or between effects of different CODECs on an image when considered over the entire frequency range. In addition to global effects, the difference signal over smaller frequency intervals enables the comparison of

---

[9] Throughout this report "frequency band" or "band" refers to a range of frequencies of the power spectrum yielded by the SIVV utility and should not be confused with compression decomposition resolutions or wavelet specifications.

effects over frequency bands that may have particular relevance to fingerprint image quality or matching, as well as quantifying and isolating changes confined to bands that specifically impact either the machine matcher or trained examiners.

3. Analysis

3.1. Normality of Metrics

Distributions of each of the metrics were examined using histograms and normal probability plots (also known as the Q-Q Plot) to determine if data follows or approximates a normal distribution. As the majority of the metrics are not normally distributed, the median is taken as the preferred measure of central tendency. Sample medians are then used to estimate the population medians for each metric.

3.2. Uncertainty of Median

For each sample median, the point-wise 95 % confidence interval is determined via a bootstrap procedure [WU1], [WU2], [WU3] in which the distribution of each metric is resampled randomly 1000 times with replacement and a median is computed for each replication. The 95 % confidence limits are then taken as the 0.025^{th} and 0.975^{th} quantile of the distribution of replicate medians for each metric. These limits are shown in tables and indicated in graphs together with the value of the median estimated from the observed distribution of each metric.

3.3. Hypothesis Testing

Hypothesis testing is performed using a non-parametric method, specifically the Wilcoxon Signed Rank Test [WILCOXON], [HOLLANDER]. Differences among the various compression processes are tested for each of three metrics. We test for differences in each of three metrics between every pair of compression processes without regard to directionality of differences (two-tailed test). We have 7 different compression conditions (i) under test yielding 21 pairwise comparisons for each of the 3 metrics (g). That is, for each pair of compression treatments, we wish to test the null hypothesis, H_0, that the median difference of pairwise measurements is zero. We reject the null hypothesis if we find the probability of its truth to be less than the Type I error rate, i.e. the probability of incorrectly rejecting the null hypothesis. Typically, the acceptable Type I error rate is set at 5%, i.e. $\alpha = 0.05$. If we are able to reject the null hypothesis, we accept the alternative hypothesis, H_1, that the median of pairwise differences in measurements is not equal to zero. Thus, for the present experiments, the null hypotheses, H_0, and alternative hypotheses, H_1, for the n=21 pairwise comparisons may be summarized as:

$$\begin{Bmatrix} H_0^{i,k,g} : \operatorname{Med}(m_i - m_k) = 0 \\ H_1^{i,k,g} : \operatorname{Med}(m_i - m_k) \neq 0 \end{Bmatrix} \Bigg|_{i,k,g}^{i=1...c; k=1...c; k \neq i;\ 1...t} \quad (4)$$

where i, k designate compression conditions $i, k=1...7$ ($k \neq i$) and g designates the metric $g = 1...3$. Pairwise comparisons are performed independently for each of the 3 metrics.

The Wilcoxon Signed Rank test examines differences between pairwise measurements as might the pairwise t-test be used to compare pairs of measurements having distributions known satisfy the assumptions of normality. Even as subsequent tables and graphs present sample medians, the comparison is made between pairs of measurements and not between distributions.

4. Results

4.1. Investigative Goal 1: Compare WSQ to JPEG 2000 in terms of fidelity to the original image

One of the fundamental questions one must ask in the transition from WSQ to JPEG 2000 is: Can JPEG 2000 operate as well as WSQ under a certain set of operational conditions? For example, how does JPEG 2000 fare against WSQ on never-before-compressed 500 ppi imagery? This is not a formal use-case that is expected in typical system operation in the United States, as the prescribed method for compression of 500 ppi imagery is currently WSQ in most systems. However, this case does provide an interesting look at how these two respective algorithms behave given the same input image. A more operationally realistic case would be the case where a 1000 ppi JPEG 2000 image is decompressed and the resulting image is recompressed with WSQ at 500 ppi, as is the case of down-sampling from 1000 ppi. Table 3 summarizes one and two pass application of WSQ and JPEG 2000 compression as described previously in 2.3.1.

Table 3 - Algorithm Combinations Tested

		Pass 1			
		No Compression	WSQ	JPEG 2000 (WSQ Rate Lock)	JPEG 2000
Pass 2	No Compression	N/A	✓ Case 1	✓ Case 2	✓ Case 3
	WSQ	N/A	✓ Case 4	N/A	✓ Case 5
	JPEG 2000 (WSQ Rate Lock)	N/A	N/A	N/A	N/A
	JPEG 2000	N/A	✓ Case 7	N/A	✓ Case 6

4.1.1. Investigative Analysis 1

Table 4 summarizes median values with 95 % confidence intervals for each of the three metrics applied to the image outputs of each of the seven processing treatments described in section 2.3. Plots of the fidelity metric values are shown below:

Table 4 – Median values of fidelity metrics for each of 7 compression treatment cases, each having N=2898 values of each of the three metrics.

PSNR (higher values denote greater fidelity to original image)

	WSQ	JP2K(WSQ rate)	JP2K	WSQ on WSQ	WSQ on JP2K	JP2K on WSQ	JP2K on JP2K
Lower CL/dB	27.703	28.704	29.265	27.690	26.984	27.371	29.251
Median/dB	27.805	28.804	29.413	27.773	27.105	27.469	29.396
Upper CL/dB	27.925	28.920	29.557	27.894	27.238	27.575	29.537

Proportion of Changed Pixels (lower values denote greater fidelity to the original image)

	WSQ	JP2K(WSQ rate)	JP2K	WSQ on WSQ	WSQ on JP2K	JP2K on WSQ	JP2K on JP2K
Lower CL/%	48.023	47.932	47.684	48.004	48.123	48.104	47.658
Median/%	48.048	47.972	47.722	48.037	48.153	48.135	47.699
Upper CL/%	48.076	48.011	47.776	48.067	48.185	48.173	47.740

SIVV Power Difference (lower values denote greater fidelity to the original image)

	WSQ	JP2K(WSQ rate)	JP2K	WSQ on WSQ	WSQ on JP2K	JP2K on WSQ	JP2K on JP2K
Lower CL/dB	1.684	3.923	3.189	1.754	1.136	2.361	3.165
Median/dB	1.754	4.021	3.294	1.832	1.213	2.463	3.273
Upper CL/dB	1.814	4.137	3.386	1.899	1.276	2.573	3.355

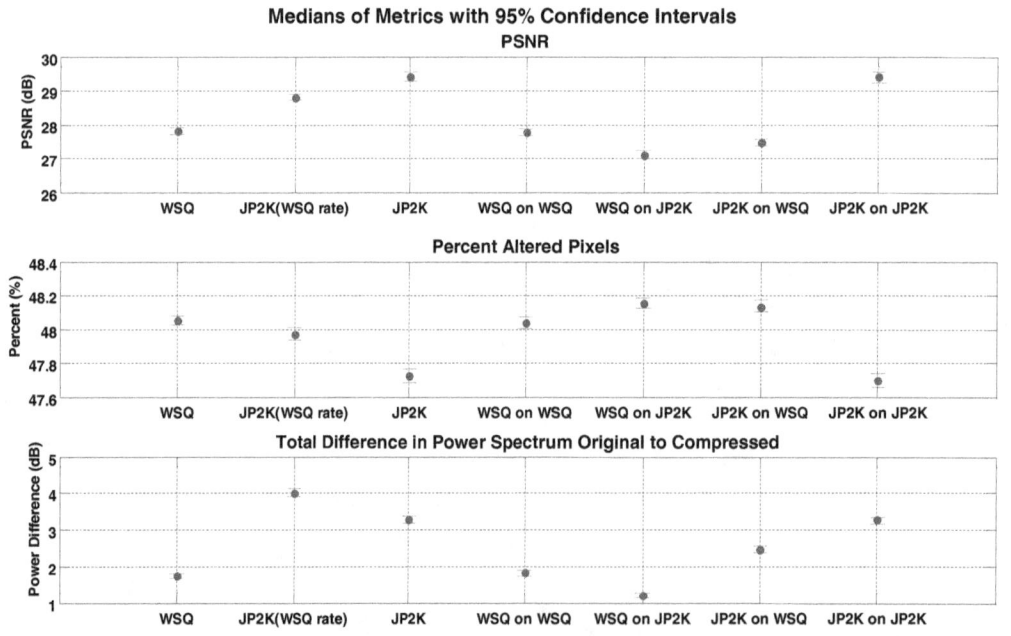

Figure 4 - Median values of three metrics with individual 95 % confidence intervals on the statistic for the seven WSQ and JPEG 2000 compression conditions examined in the present investigation for each of the three metrics.

Several observations may be made from the examination of Table 4 and Figure 4. First, it appears that PSNR and Proportion of Altered Pixels are essentially inversely related to one another, suggesting that the magnitude of the change in images indicated by the PSNR is distributed with some regularity among the altered pixels. Thus, the tabulation of altered pixels carries essentially the same information as the PSNR in this case.

Second, PSNR of JPEG 2000 compressed images is consistently greater than that for WSQ. This is the case for both JPEG 2000 images compressed at rates matching those of WSQ outputs as well as JPEG 2000 images compressed at the single fixed rate of 0.55 bpp. PSNR of JPEG 2000 at the fixed rate of 0.55 bpp exceeds that where the compression rate is adjusted to match that of the WSQ output. This is consistent with the often higher compression observed when the JPEG 2000 rate is selected to match that of the WSQ output that itself varies considerably about a specified target compression rate (see Appendix A.)

Third, among the multiple compression cases, PSNR for two cycles of WSQ is very close to that for the single stage of WSQ application and PSNR for two cycles of JPEG 2000 is close to that for the single stage of compression using this CODEC. The lowest median PSNR values are found through compression/decompression with one CODEC followed by recompression with the other CODEC, with slightly higher PSNR when JPEG 2000 is applied to images having first been subjected to processing with WSQ.

Thus, noting the caveat that PSNR is not necessarily the most appropriate measure of image fidelity [GIROD], PSNR tends to indicate that JPEG 2000 compression has less impact on pixel values than does WSQ compression.

The SIVV spectral signal however tells a somewhat different story. The SIVV method summarizes the 2D power spectrum of the fingerprint image by integrating the polar transform of the 2D spectrum over angle [LIBERT]. As a summary of the image spectrum, it enables visualization of the frequency structure of the image. Comparison of the SIVV signals of compressed and non-compressed images exhibits the alteration of the power distribution caused by the compression. Figure 5 shows a typical comparison of WSQ and JPEG 2000 SIVV signals against that of the non-compressed version of the specimen image.

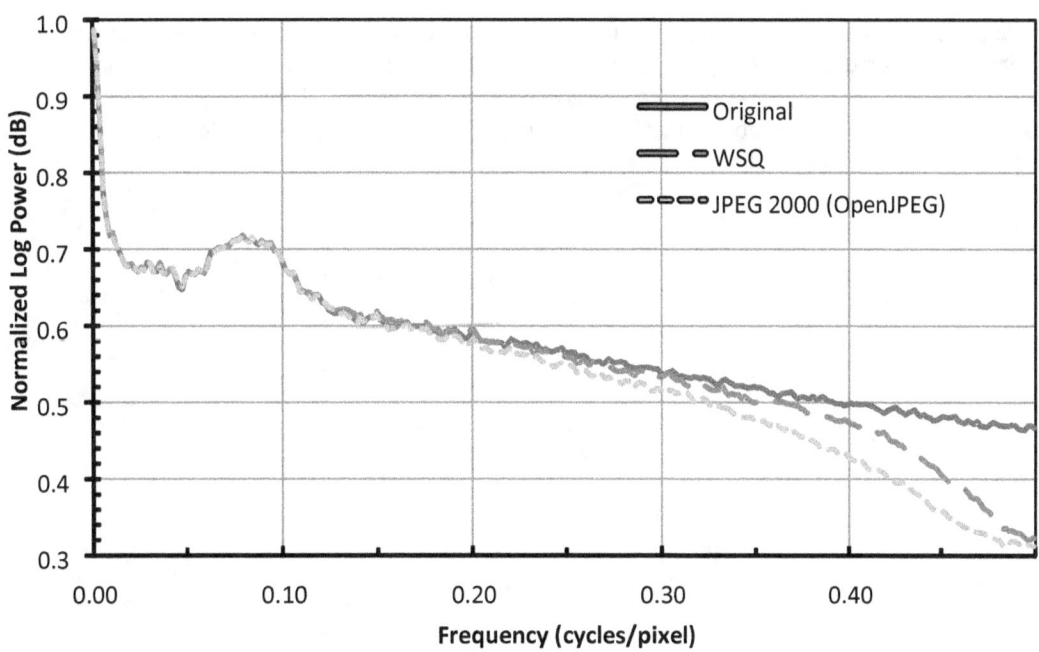

Figure 5 - Example of typical behavioral differences between JPEG 2000 and WSQ (15:1 compression)

This figure typifies the comparative SIVV representation of WSQ and JPEG 2000. WSQ tends to better match the spectrum of the non-compressed image over most of the frequency range, finally taking a rather steep fall-off in the high frequency. By contrast, JPEG 2000 tends to depart from the non-compressed spectrum at lower frequencies and continues to lose power, almost linearly in many cases, into the high frequency band where the slope of the spectrum steepens. The observed behavior of the WSQ algorithm is not accidental. The WSQ wavelet decomposition structure was designed to preserve as much as possible of the frequency information over much of the spectrum where identification features reside, electing intentionally to discard much of the high frequency content [HOPPER].

To compare differences in SIVV signals statistically, we compute the median of the sums of differences in the power spectra of non-compressed original images and corresponding processed versions of those images. Referring to these data in the lower plot of Figure 4, we see that for the most part the PSNR relationship is reversed [10]. With respect to the single compression cycles, we find WSQ to show lower power loss compared to the two JPEG 2000 cases. JPEG 2000 images compressed at rates matched to the WSQ outputs show the greatest loss in power at more than 4 dB from the non-compressed original, fairly considerable for a median value. The lower power loss for JPEG 2000 compressed at the single rate of 0.55 bpp, in contrast to that of JPEG 2000 at the WSQ compression rate, reflects the generally lower and more predictable compression rate behavior of JPEG 2000 (discussed in Appendix A). Still, the median power loss of nearly 1.5 dB greater than that of WSQ is notable.

The power loss results show us that 2 cycles of either CODEC show essentially the same loss as a single cycle, with WSQ showing the lower loss in spectral power. Power loss with application of JPEG 2000 on the decoded output of the WSQ CODEC is intermediate to that of WSQ or JPEG 2000. The target compression rate specified to the JPEG 2000 CODEC in this case is the single 0.55 bpp value. An unexpected result is that when JPEG 2000 is applied to the decoded outputs of the WSQ CODEC the resultant spectra are closer to those of the non-compressed original images than either of the single cycle applications of either CODEC. Definitely determining the cause of this may be the subject of additional investigation, but we offer the preliminary hypothesis that the interaction of the wavelet decomposition of the different CODECs may introduce noise, harmonics, or spurious contrast (see Appendix B) features that appear as additional undifferentiated power in the summed SIVV difference signals. Indeed, observation of SIVV signal comparison plots displayed during processing shows a spurious increase in power even in high frequency bands.

4.1.1.1. Hypothesis Testing

We use the Wilcoxon Signed Rank test to evaluate the significance of differences among the 7 compression processing treatments. As indicated in (3.3) we compare the various compression processes in a pairwise fashion. Table 5 summarizes the pairwise comparison of every combination of compression treatments for each of the three metrics. As is apparent in inspecting Table 5, in all the null hypothesis is rejected with a probability (p) of a Type I Error much less than even the error threshold of $\alpha=0.05$ in most cases.

---

[10] Note here that smaller difference between spectra of non-compressed and compressed generally indicates lower distortion, i.e., higher fidelity in contrast to PSNR where higher values are preferred.

Table 5 – Simultaneous Hypothesis Tests of Differences Between Processing Treatments (α=0.05)

Comparison #	Process 1	Process 2	PSNR (dB)			Proportion Changed Pixels			Power Difference		
			Median 1	Median 2	p	Median 1	Median 2	p	Median 1	Median 2	p
1	WSQ	JP2K(WSQ rate)	27.805	28.804	0.0000	48.048%	47.972%	0.0000	1.7545	4.0213	0.0000
2	WSQ	JP2K	27.805	29.413	0.0000	48.048%	47.722%	0.0000	1.7545	3.2944	0.0000
3	WSQ	WSQ on JP2K	27.805	27.105	0.0000	48.048%	48.153%	0.0000	1.7545	1.2129	0.0000
4	WSQ	JP2K on WSQ	27.805	27.469	0.0000	48.048%	48.135%	0.0000	1.7545	2.4634	0.0000
5	WSQ	WSQ on WSQ	27.805	27.773	0.0000	48.048%	48.037%	0.0225	1.7545	1.8315	0.0000
6	WSQ	JP2K on JP2K	27.805	29.396	0.0000	48.048%	47.699%	0.0000	1.7545	3.2728	0.0000
7	JP2K(WSQ rate)	JP2K	28.804	29.413	0.0000	47.972%	47.722%	0.0000	4.0213	3.2944	0.0000
8	JP2K(WSQ rate)	WSQ on JP2K	28.804	27.105	0.0000	47.972%	48.153%	0.0000	4.0213	1.2129	0.0000
9	JP2K(WSQ rate)	JP2K on WSQ	28.804	27.469	0.0000	47.972%	48.135%	0.0000	4.0213	2.4634	0.0000
10	JP2K(WSQ rate)	WSQ on WSQ	28.804	27.773	0.0000	47.972%	48.037%	0.0000	4.0213	1.8315	0.0000
11	JP2K(WSQ rate)	JP2K on JP2K	28.804	29.396	0.0000	47.972%	47.699%	0.0000	4.0213	3.2728	0.0000
12	JP2K	WSQ on JP2K	29.413	27.105	0.0000	47.722%	48.153%	0.0000	3.2944	1.2129	0.0000
13	JP2K	JP2K on WSQ	29.413	27.469	0.0000	47.722%	48.135%	0.0000	3.2944	2.4634	0.0000
14	JP2K	WSQ on WSQ	29.413	27.773	0.0000	47.722%	48.037%	0.0000	3.2944	1.8315	0.0000
15	JP2K	JP2K on JP2K	29.413	29.396	0.0000	47.722%	47.699%	0.0000	3.2944	3.2728	0.0000
16	WSQ on JP2K	JP2K on WSQ	27.105	27.469	0.0000	48.153%	48.135%	0.0025	1.2129	2.4634	0.0000
17	WSQ on JP2K	WSQ on WSQ	27.105	27.773	0.0000	48.153%	48.037%	0.0000	1.2129	1.8315	0.0000
18	WSQ on JP2K	JP2K on JP2K	27.105	29.396	0.0000	48.153%	47.699%	0.0000	1.2129	3.2728	0.0000
19	JP2K on WSQ	WSQ on WSQ	27.469	27.773	0.0000	48.135%	48.037%	0.0000	2.4634	1.8315	0.0000
20	JP2K on WSQ	JP2K on JP2K	27.469	29.396	0.0000	48.135%	47.699%	0.0000	2.4634	3.2728	0.0000
21	WSQ on WSQ	JP2K on JP2K	27.773	29.396	0.0000	48.037%	47.699%	0.0000	1.8315	3.2728	0.0000

Whereas median total difference in spectral power is considered as one of our three primary metrics, useful information is available in examining the spectral difference in various frequency bands of the SIVV signal. In Figure 6 we see the median differences in the summed power of the SIVV signals on a band-by-band basis. As indicated previously, each frequency band represents an interval of 0.1 cycles/pixel within the overall frequency range of 0.0 cycles/pixel to 0.5 cycles/pixel. Also, shown are the medians of total summed differences for each of the compression cases. Single- and dual-cycle compression with WSQ track closely together and show relatively little loss in power, i.e., small median differences in SIVV signal, over most of the frequency range (Bands 1 – 4), with considerable loss in Band 5. Notably, the WSQ and JPEG 2000 on WSQ cases show a small negative difference apparent in Bands 1 and 2. As indicated via the analysis presented in Appendix B, this could result from an increase in contrast resulting from the WSQ compression process though additional analysis would be required to confirm that WSQ increases the contrast of the image.

Figure 6 shows substantial power loss over Bands 3 – 5 with JPEG 2000 compression, exhibiting virtually superimposed plot lines. This appears consistent with Lepley's [MTR] observation that visual examination of images exhibited a "softening" of image features with JPEG 2000 compression relative to WSQ processing. Moreover, as the NIST SIVV metric was derived from frequency spectrum properties elaborated by Nill [NILL1], SIVV comparisons described above are similar to those using the IQM metric [NILL4] for which Lepley reports higher quality for WSQ in contrast to JPEG 2000 images.

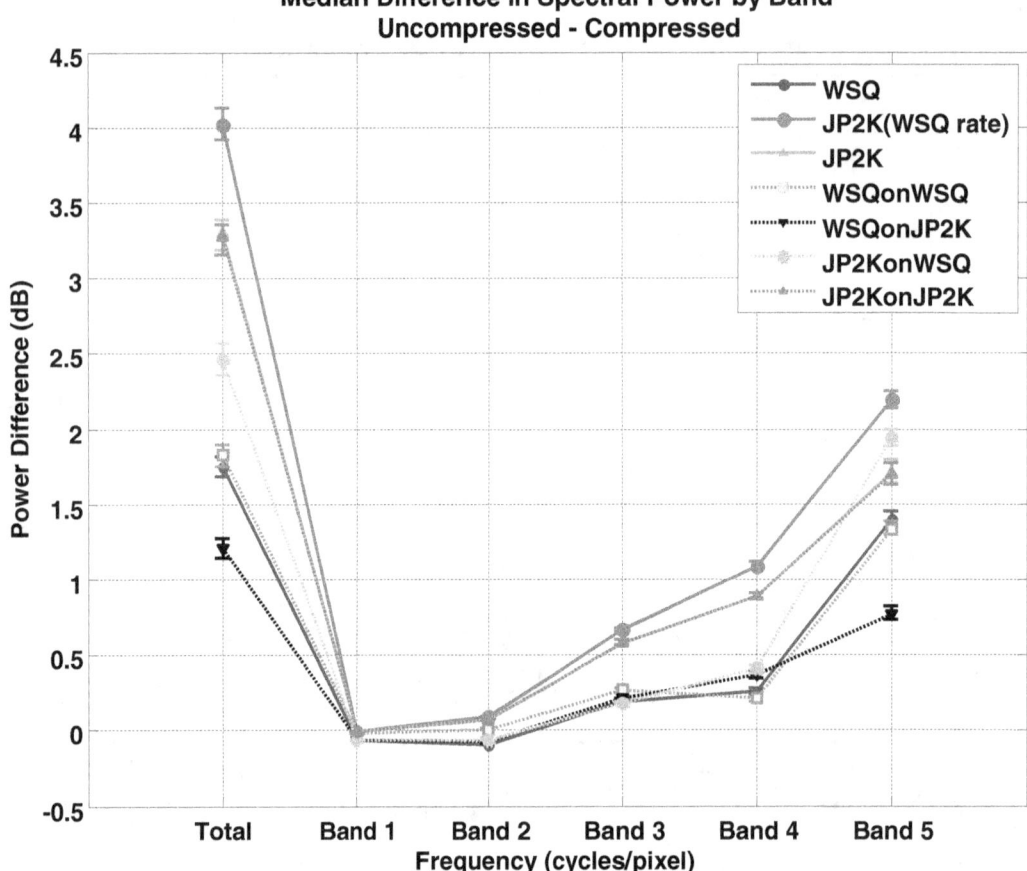

Figure 6 - Median values of three metrics with 95 % confidence intervals on the statistic for seven WSQ and JPEG 2000 compression conditions examined in the present investigation. (Note that larger power differences indicate lower fidelity to non-compressed original image).

4.1.1.2. Relation to Expert Examiner Judgment

A test of human subjective assessment of image quality was not conducted for the 500 ppi images used in the present study. However, NIST conducted a large scale test [NISTIR 7778] of expert fingerprint examiner response to pairs of 1000 ppi fingerprint images including a subset of images drawn from the SD27A dataset that are also used in the present study. Three expert fingerprint examiners were presented with a non-compressed fingerprint image paired with either a matching or non-matching fingerprint, either also non-compressed or compressed using JPEG 2000 at one of 14 compression ratios ranging from 2:1 to 38:1. Each image pair was presented at each of the 14 compression ratios at random intervals during the study. Examiners were asked to first determine if the pair represented a "match," i.e., prints from the same finger of an individual. Examiners were then to classify the "quality" of the poorer of the two prints with respect to preservation of image features essential to or useful for identification. The classification used a 4-point scale representing conditions[11] describing the relative quality of one image as compared to the other. The incidence of rating 4 was judged too infrequent for consideration here.

The signal metrics described above for the 500 ppi images used in the present study are applied to the 1000 ppi images used in the examiner study. These 1000 ppi images correspond to a subset of the 500 ppi image used in the present study. The ratings 1 – 3 are used to group the quantitative measurements PSNR and the SIVV median power difference metrics made on the 1000 ppi images.

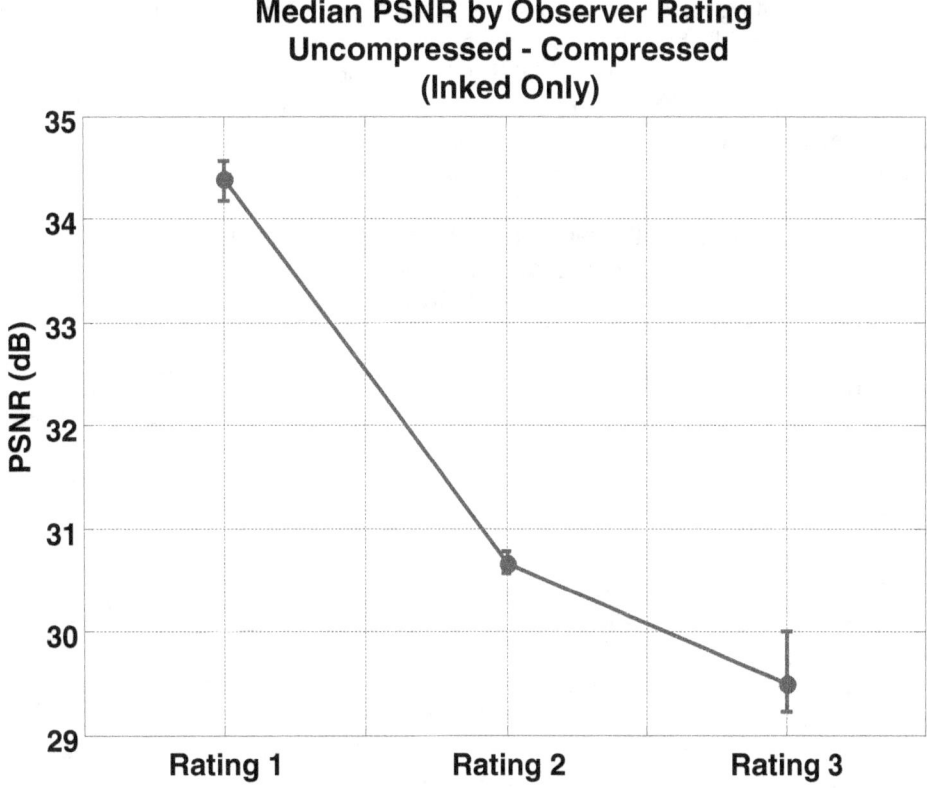

Figure 7 - Median PSNR and 95 % confidence intervals for 1000 ppi, inked fingerprint images classified by expert examiners into each of the degradation categories 1 – 3. Only matching pairs of non-compressed and compressed versions of the same image are included here; i.e. same subject, same finger.

While not strictly comparable to measurements of the 500 ppi images used in the present study, Figure 7 and Figure 8 show respectively PSNR and SIVV difference results for images classified by examiners as having visual quality 1, 2, and 3 with respect to the non-compressed original. (We note that category 4 is omitted due to the small number of images assigned that rating, noting that at least at 1000 ppi, the Galton features tolerate considerable compression.)

---

[11] The ratings and conditions are as follows:
(1) No apparent image quality degradation and the quality of Level II(2) and Level III(3) detail in either image should not cause any difficulty in reaching a conclusive decision of identification or exclusion.
(2) A noticeable degradation in the quality of Level II(2) or Level III(3) detail in either image, but not enough to have a negative impact on reaching a conclusive decision of identification or exclusion, though the amount of time to reach a decision may increase.
(3) Level III(3) detail quality diminished in either image to the extent that a Level III(3) identification is questionable or not possible, and/or is significantly more difficult.
(4) Level II(2) detail quality diminished in either image to the extent that a Level II(2) identification becomes questionable or not possible, and/or is significantly more difficult.

Figure 7 shows a marked drop in median PSNR between rating 1 (little or no degradation) and rating 2 (observable degradation, but no loss of features used for identification). A further drop in PSNR is found for images assigned rating 3 indicating loss of non-Galton identification features, such as pores and ridge shape [MALTONI], [JAIN]. Confidence intervals (95 %) on each of the medians are indicated via error bars[12].

Figure 8 shows, for images assigned each of the ratings 1 – 3, the median difference in spectral power both over the entire spectrum as well as for each of the 5 frequency bands as described previously. The loss in total spectral power is least among images assigned rating 1 and marked for both ratings 2 and 3. The breakout by band of the power loss indicates that the features in the middle to high frequencies are largely those involved in identification. The power losses in those channels increase exponentially toward the high frequency band. Interestingly, there is a substantial difference in power loss between images at rating 1 judged to be not degraded at all and those at rating 2, judged to be degraded but to have retained all useful detail for identification. Yet, a comparatively small additional power loss in the mid to high frequencies correlates with loss of level 3 features useful for identification.

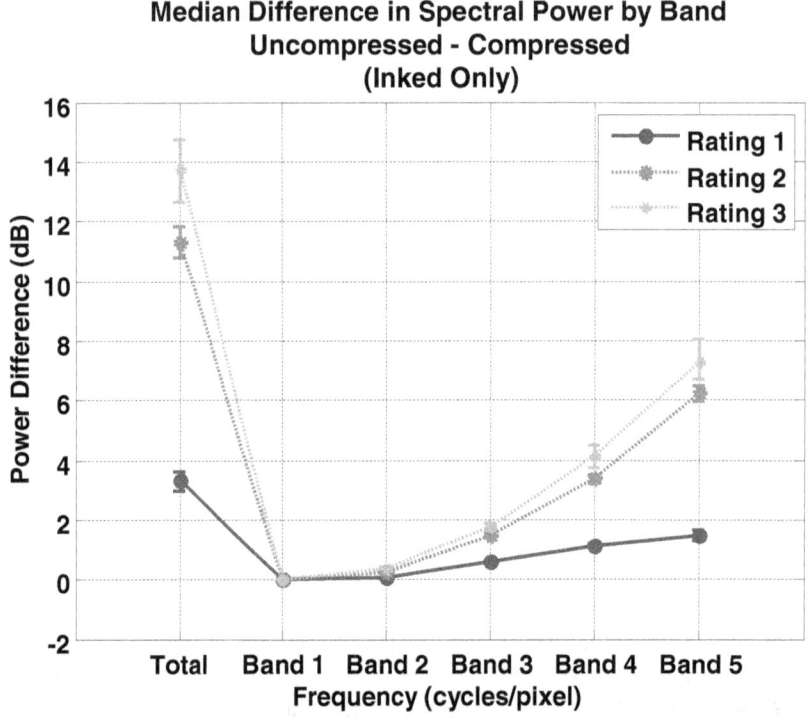

Figure 8 - Median power difference by frequency band with individual 95 % confidence intervals for 1000 ppi fingerprints subjected to various levels of JPEG 2000 compression grouped by ratings assigned by expert fingerprint examiners relative to degradation of features useful for identification. Total power difference is shown as well as sum of difference over each of 5 frequency bands from 0.0 cycles/pixel to 0.5 cycles/pixel.

4.1.2. Investigative Results 1

Given the traditional definition of the number of altered pixels when comparing a processed to the original, Table 4 and Figure 4 show that JPEG 2000 exhibits better performance characteristics than WSQ by altering fewer pixels. WSQ on the other hand demonstrates better linearity with the original image signal much further into the frequency spectrum as demonstrated by Figure 5 followed by a rapid drop-off in the higher frequency portions of the image. It is also observed that applying the same algorithm twice does not have an impact nearly as significant as mixing algorithms. For example, two passes of WSQ yield PSNR and Percent Altered Pixel values comparable to a single pass

---

[12] The width of the confidence intervals in this case are indicative of the number of images classified in each of the three categories, i.e. most cases classified with rating 2, next with rating 1, and the least assigned rating 3.

of WSQ, as is the case with JPEG 2000. However, applying WSQ to a JPEG 2000 image, or vice versa, causes a more marked pixel alteration in the image.

4.2. Investigative Goal 2: Examine effects of multiple compression cycles

In the preceding section, we observed the relative effects of 2 cycle compression using WSQ and JPEG 2000 as well as mixed application of the two CODECs. Here we examine the stability of the CODECs under the extreme repetition of encoding and decoding each of a random sample of fingerprint images.

4.2.1. Investigative Analysis 2

Figure 9 illustrates the stability of JPEG 2000 over multiple compression cycles. The median PSNR of a random sample of 200 fingerprint images remains constant over 200 cycles of compression, decompression, and recompression. By contrast, WSQ shows decay of PSNR over the same number of cycles to a minimum PSNR level.

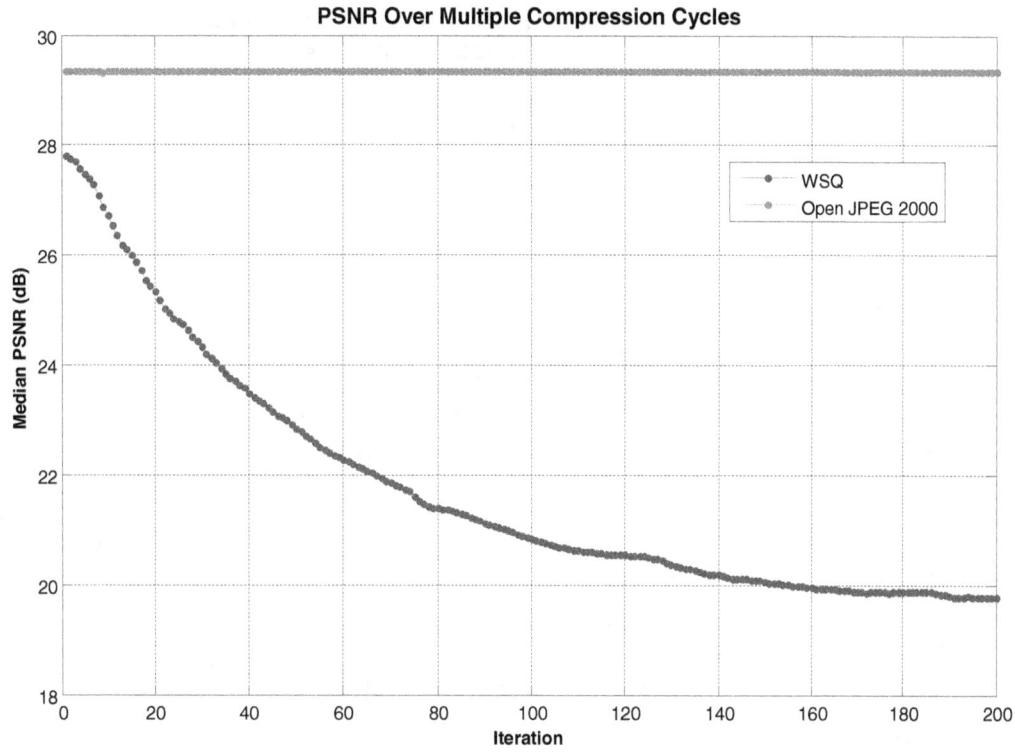

Figure 9 - Median Peak Signal to Noise Ratio over 200 compression cycles applied to each of 200 fingerprint images. In contrast with WSQ, JPEG2000 exhibits remarkable stability even at such extremes of repeat compression/decompression. (In each case, PSNR compares the multiple compression cycle output to the non-compressed original).

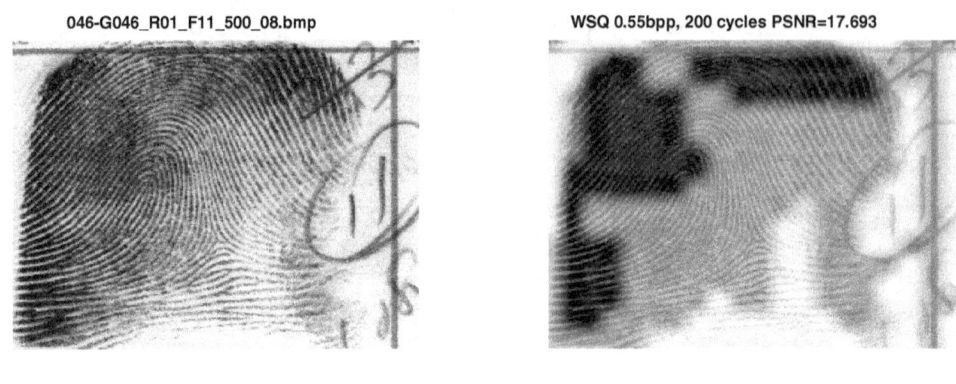

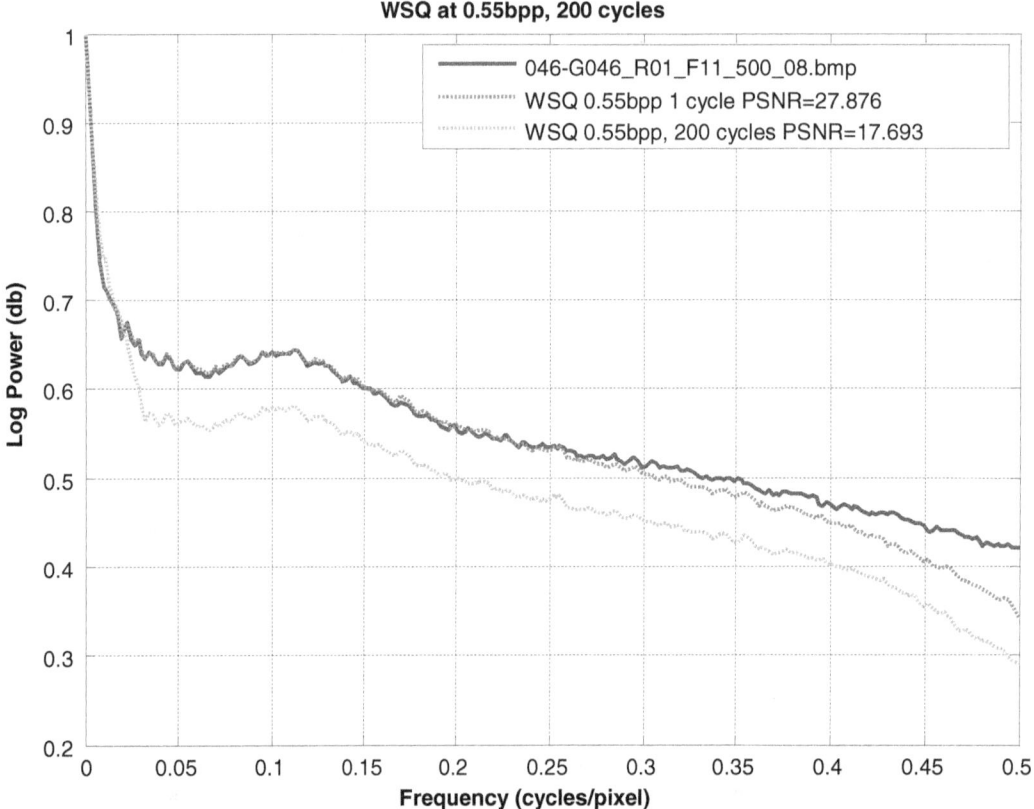

Figure 10 – Images at upper left and right show respectively one of the non-compressed original images used in the experiment and the result of 200 cycles of compression/decompression with the WSQ CODEC. The lower graph shows the spectra of the non-compressed image, the result of a single cycle of WSQ compression, and that of 200 cycles. PSNR values (in decibel units) appear in the legend. The spectral pattern is similar for 1 cycle and 200 cycles but power at all frequencies is reduced.

Figure 10 shows the effects of multiple WSQ compression cycles on an example image among those sampled. PSNR is reduced substantially along with contrast, which is evident in the uniform drop in power over the entire frequency range relative to the spectrum shown for a single compression cycle at the target compression rate of 0.55 bpp. Appendix B provides some discussion of the effect of contrast on the power spectrum and provides a contrast metric as the ratio of the gray level standard deviation to the mean pixel value. In the case of the example considered here, the contrast measure, C, is 0.62 for the non-compressed image and 0.62 and 0.59 respectively for the single and multiple compression cycles.

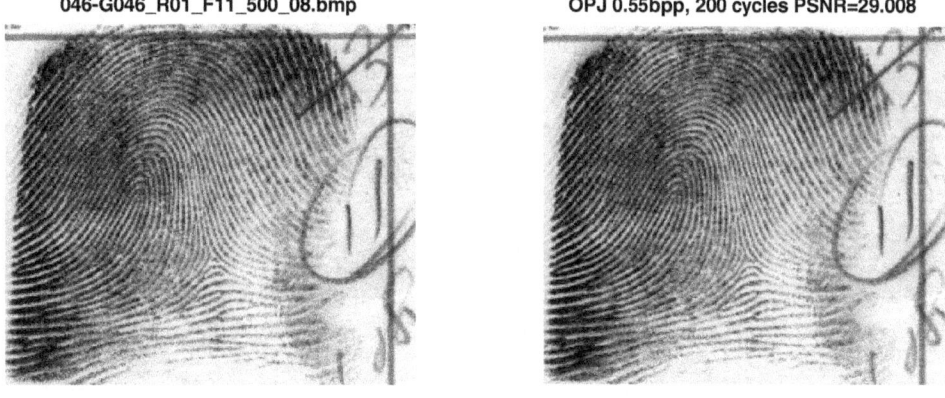

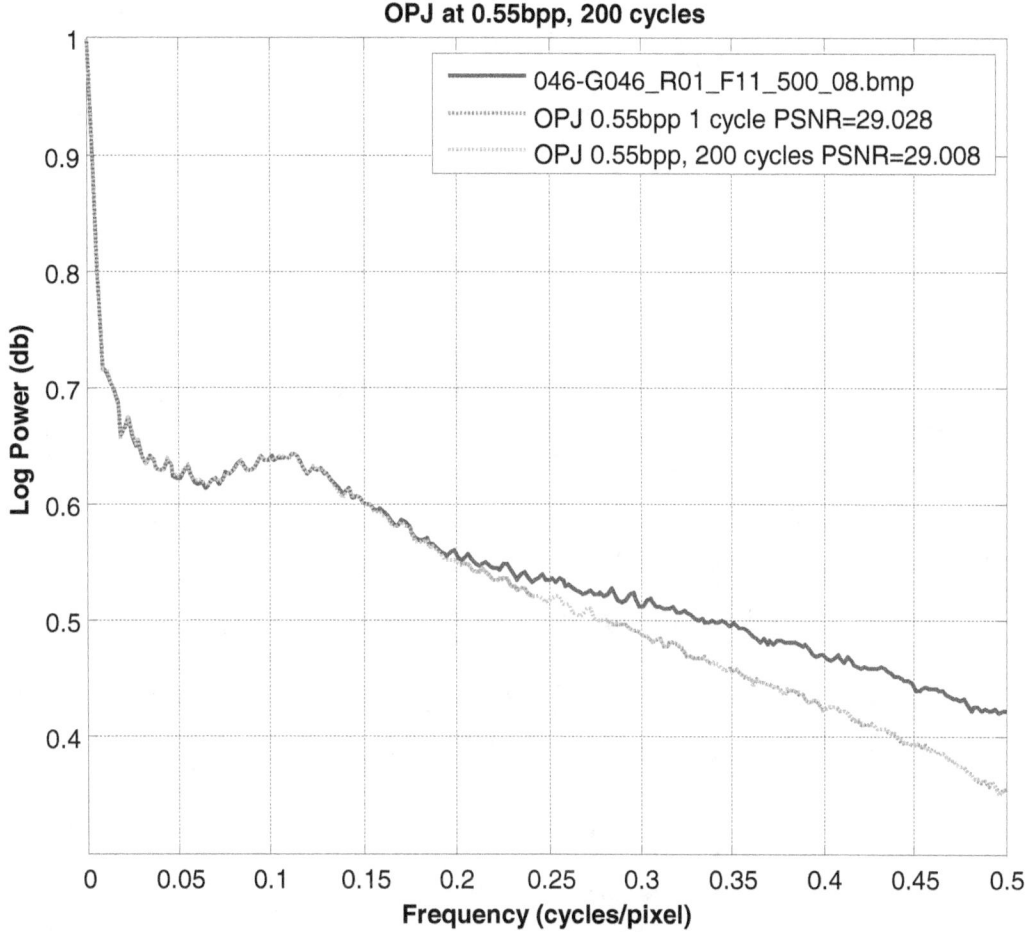

Figure 11 – Images at upper left and right are respectively one of the non-compressed original images and that resulting from 200 cycles of compression/decompression using the JPEG 2000 CODEC. The lower plot shows that while spectral power degradation is evident as early as in the middle of the frequency range and increases toward the high frequency, this degradation is relatively stable even up to 200 cycles of compression and is virtually identical to that after only a single cycle of compression using this CODEC. PSNR difference is negligible as well.

Figure 11 shows the effects of multiple compression cycles on a non-compressed example image using the JPEG 2000 CODEC. We see that PSNR and spectrum remain virtually the same between the result of a single compression cycle and that of 200 cycles. The contrast is 0.62 for all three images.

4.2.2. Investigative Results 2

The data collected in this analysis phase demonstrates that JPEG 2000 provides better resiliency in situations where multiple compression/decompression cycles are expected. While operationally, only one compression cycle is expected, there are certain cases where multiple cycles may occur such as segmentation; where multi-finger images are cropped and individual fingers are separated from the larger image and recompressed. Any such case expected to incur more than one compression cycle may benefit from the stability of JPEG 2000.

5. Conclusions

5.1. Fidelity: Single Compression Cycle

With respect to fidelity we find the results to be somewhat ambiguous. PSNR and similar measures such as the proportion of changed pixels tend to show less distortion with JPEG 2000 than with WSQ. Yet changes in the image spectrum with WSQ are less than with JPEG 2000, particularly in the middle of the frequency range, most likely to contain fingerprint features used for identification. However, comparisons evaluated in the present study involve 500 ppi imagery, not generally processed using the JPEG 2000 CODEC. WSQ will not be applied to 1000 ppi and higher sample rate imagery; hence the distinction may be moot for the most part. However, the conclusion that JPEG 2000 is not optimally tuned for fingerprint imagery may advise conservatism with regard to pushing high compression rates with the new CODEC, particularly when one expects to extract a 500 ppi image from the compressed 1000 ppi code stream, a process quite easily accomplished with JPEG 2000 Part 1 implementations.

Regarding the apparent conflict between PSNR results and the power spectral comparison, we note that JPEG 2000 employs a rate distortion optimization logic designed to optimize mean squared error of wavelet packets within the constraint of a targeted output bit rate or compression ratio [TAUBMAN1], [TAUBMAN2]. In JPEG2000 Part 1, under consideration here, options for such optimization are somewhat limited, but Part 2 of the standard provides for much greater latitude for control of the wavelet decomposition and optimization, including some interesting provisions to improve the JPEG 2000 compression of fingerprint images [STÜTZ].

5.2. Fidelity: Two compression cycles each using either WSQ or JPEG 2000 CODECs

Either CODEC appears to behave satisfactorily for up to two cycles of encoding and coding. As this is likely to occur in practice, the relatively low loss in quality observed in the present experiments is encouraging.

5.3. Fidelity: Two compression cycles using dissimilar encoders

Whereas both WSQ and JPEG 2000 CODECs appear to perform well with up to 2 cycles of encoding and decoding, the same is not true when the CODECs are mixed. Both cases show reduction in fidelity as measured by PSNR and proportion of changed pixels. Spectral comparison is somewhat ambiguous, but does indicate that some modifications are being made to the frequency structure of the image with mixed CODEC processing.

5.4. Effects of compression/decompression over extreme numbers of multiple cycles

For this scenario, JPEG 2000 exhibits remarkable stability over multiple compression cycles. PSNR and power spectrum are altered only negligibly even over 200 cycles of compression, decompression, and recompression. WSQ, by contrast, shows considerable degradation over even a few cycles and continues degradation asymptotically to some minimum level of fidelity.

6. Future Work

The present study aimed mainly at exploring signal fidelity of images processed by the WSQ and JPEG 2000 CODECs under operational and extreme scenarios of application. Ultimately the utility of either CODEC remains a function of how compression processing might affect the ability to use the fingerprint for identification. Accordingly, experiments are planned to compare the processed images using an Automated Fingerprint Identification System (AFIS) matcher and to have trained fingerprint examiners provide comparative visual assessments of the variously compressed fingerprint images using a variation on the experimental method detailed in [NISTIR7778]. Additional experimentation may follow with some of the proposed JPEG2000 Part 2 modifications to the wavelet decomposition and rate distortion optimization of JPEG 2000 to better match the frequency preservation characteristics of the WSQ codec.

References

Publications and Reports

BRADLEY1	Bradley, Jonathan N., Brislawn, Christopher M., Hopper, Thomas. (1993). "FBI wavelet/scalar quantization standard for grayscale fingerprint image compression", SPIE Conference on *Visual Information Processing II*, eds. Huck, Friedrich O., Juday, Richard D. pp. 293-304.
BRADLEY2	Bradley, J. N., Brislawn, C. M. (1994). "The wavelet/scalar quantization compression standard for digital fingerprint images". IEEE International Symposium on Circuits and Systems, 1994. ISCAS '94., 1994. Vol 3. pp. 205-208.
BRISLAWN1	Brislawn, Christopher M. (1996). "Wavelet Scalar Quantization Compression Standard for Fingerprint Images". *Proceeding Conference on Signal Image Processing and Applications*.
BRISLAWN2	Brislawn, Christopher M. (2002). The FBI Fingerprint Image Compression Standard. http://www.c3.lanl.gov/~brislawn/FBI/FBI.html, updated 25 June 2001 (accessed 01/31/2014).
CHAI	Chai, D., Bouzerdoum, A. (2001). "JPEG 2000 image compression: an overview". *Proceedings of The Seventh Australian and New Zealand 2001 Intelligent Information Systems Conference*, pp. 237-241.
CHAMBERS	Chambers, John; William Cleveland, Beat Kleiner, and Paul Tukey (1983). Graphical Methods for Data Analysis. Wadsworth
FBI	FBI CRIMINAL JUSTICE INFORMATION SERVICES (CJIS) (1997). *WSQ GRAYSCALE FINGERPRINT IMAGE COMPRESSION SPECIFICATION. IAFIS-IC-0110(V3)*. December 19, 1997.
FIG-VILL	Figueroa-Villanueva, Miguel A., Nalini K. Ratha, Ruud M. Bolle. (2003). "A comparative performance analysis of JPEG 2000 vs. WSQ for fingerprint image compression". *Proceedings of the 4th international conference on Audio- and video-based biometric person authentication*. pp. 385-392.
FITZPATRICK	Fitzpatrick, M. Et al. 1994, "WSQ Compression / Decompression Algorithm Test Report", IAI Annual Conference.
FUNK	Funk, W., Arnold, M., Busch, C., Munde, A. (2005). "Evaluation of image compression algorithms for fingerprint and face recognition systems". *Proceedings from the Sixth Annual IEEE SMC Information Assurance Workshop, 2005. IAW '05*. pp. 72-78.
GALTON	Galton, F. (2005). *Finger prints*. Mineola, NY: Dover Publications. (Original work published 1892)
GIROD	Girod, Bernd (1993). "What's wrong with mean-squared error?" *Digital images and human vision*, Andrew B. Watson, ed. MIT Press, pages 207-220.
HOLLANDER	Hollander, M., Wolfe, D. A. (1999). Non-Parametric Statistical Methods, 2nd Ed, John Wiley & Sons:NY, 787 pages.
HOPPER	Hopper, T., Preston, F. (1992). "Compression of grey-scale fingerprint images", *Proceedings of Data Compression Conference, 1992. DCC '92*. pages 309-318.
HUYNH	Huynh-Thu, Q., Ghanbari, M. (2008). "Scope of validity of PSNR in image/video quality assessment". *Electronics Letters*, Vol. 4, No. 13, pp. 800-801.
JAIN	Jail, A., "Pores and Ridges: High-Resolution Fingerprint Matching Using Level 3 Features", IEEE Transactions on Pattern Analysis and Machine Intelligence, Vol. 29, No. 1, January 2007.
KUMAR	Kumar, Satish. (2001). An *Introduction to Image Compression*. http://www.debugmode.com/imagecmp/
LIBERT	Libert, J.M.; Grantham, J.; Orandi, S. "A 1D Spectral Image Validation/Verification Metric for Fingerprints". NISTIR 7599, August 19, 2009. http://www.nist.gov/customcf/get_pdf.cfm?pub_id=903078
LIKERT	Likert, R. (1932). A *Technique for the Measurement of Attitudes*, Archives of Psychology 140, 55.

MALTONI	Maltoni, D., Maio, D., Jain, A. K., and Prabhakar, S. Handbook of Fingerprint Recognition (2nd ed.). Springer-Verlag: London, 2009, pp. 494.
MAS-KAM	Mascher-kampfer, A., Stögner, Herbert, Uhl, Andreas. (2007). "Comparison of Compression Algorithms: Impact on Fingerprint and Face Recognition Accuracy". *Visual Computing and Image Processing VCIP 07, Proceedings of SPIE.*
MILL	K. Millard, Developments on Automatic Fingerprint Recognition, 1983 International Carnahan Conference on Security Technology, pp.173-178, 4 October 1983.
MTR	"JPEG 2000 and WSQ Image Compression Interoperability." Lepley, M. A. MITRE Technical Report, MTR 00B0000063, February 2001 http://www.mitre.org/work/tech_papers/tech_papers_01/lepley_JPEG2000/index.html
NILL1	Norman B. Nill and Brian Bouzas, (1992). "Objective image quality measure derived from digital image power spectra", Opt. Eng. 31, 813 (1992); doi:10.1117/12.56114
NILL2	Nill, Norman. (1979). "Contrast effect on imagery power spectra," Applied Optics, Vol. 18, No. 13, pp. 2147-2151.
NILL3	Nill, Norman. (2007) IQF (image Quality of Fingerprint) Software Application. MITRE Technical Report 070053.
NILL4	Nill, Norman. (2008). Image Quality Metric (IQM). http://www.mitre.org/tech/mtf/
NIST1	National Institute of Standards and Technology. *Summary of NIST Patriot Act Recommendations.* Gaithersburg, MD. Retrieved January 4, 2007 from http://www.itl.nist.gov/iad/894.03/pact/NIST_PACT_REC.pdf
NIST2	"NIST Biometric Image Software". Http://Fingerprint.nist.gov/NFIS/. Retrieved 2011-01-12.
NISTIR7778	Orandi, S., Libert, J. M., Grantham, J. D., Ko, K., Wood, S.S., Wu, J. *Effects of JPEG 2000 Image Compression on 1000ppi Fingerprint Imagery,* NIST Interagency Report 7778, National Institutes of Standards and Technology, Gaithersburg, MD. April 11, 2011, 72 pages.
OPENJPEG	"OpenJPEG library : an open source JPEG 2000 CODEC". Http://www.openjpeg.org/index.php?menu=news. Retrieved 2011-01-12.
RAJAN	Rajan L. Joshi, Majid Rabbani and Margaret A. Lepley, "Comparison of multiple compression cycle performance for JPEG and JPEG 2000", Proc. SPIE 4115, 492 (2000)
SD27	M.D. Garris & R.M. McCabe, "NIST Special Database 27: Fingerprint Minutiae from Latent and Matching Tenprint Images," NIST Technical Report NISTIR 6534 & CD-ROM, June 2000.
SHAPIRO	Shapiro, S. S.; Wilk, M. B. (1965). "An analysis of variance test for normality (complete samples)". *Biometrika,* 52 (3-4): 591–611. doi:10.1093/biomet/52.3-4.591. JSTOR 2333709MR205384.
STÜTZ	Stütz, T., Mühlbacher, B., Uhl, A. (2010). "Best wavelet packet bases in a JPEG2000 rate-distortion sense: The impact of header data." IEEE International Conference on Multimedia and Expo (ICME), 2010. pp 19-24.
TAUBMAN1	Taubman, D. And Marcellin. (2002). *JPEG 2000: Image compression fundamentals, standards and practice.* Boston, Kluwer Academic Publishers, 795 pages.
TAUBMAN2	Taubman, DS. & Marcellin, MW. (2002). "JPEG 2000: Standard for interactive imaging". *Proceedings of the IEEE,* vol 90(8), pp. 1336 - 1357.
WILCOXON	Wilcoxon, Frank (Dec 1945). "Individual comparisons by ranking methods". *Biometrics Bulletin* 1 (6): 80–83.
WU1	Wu, Jin Chu, Alvin F. Martin, and Raghu N. Kacker. Measures, uncertainties, and significance test in operational ROC analysis, Journal of Research of the National Institute of Standards and Technology, 116(1), 517-537, (2011).
WU2	Wu, Jin Chu. Studies of Operational Measurement of ROC Curve on Large Fingerprint Data Sets Using Two-Sample Bootstrap. NISTIR 7449, U.S. Department of Commerce, National Institute of Standards and Technology, September 2007, 25 pages.

| WU3 | Wu, Jin Chu. Operational Measures and Accuracies of ROC Curve on Large Fingerprint Data Sets. NISTIR 7495, U.S. Department of Commerce, National Institute of Standards and Technology, May 2008, 23 pages. |

Standards

AN27	NIST Special Publication 500-271: American National Standard for Information Systems — *Data Format for the Interchange of Fingerprint, Facial, & Other Biometric Information – Part 1*. (ANSI/NIST ITL 1-2007). Approved April 20, 2007.
JPEG	International Telecommunications Union (ITU). "T.81 : Information technology – Digital compression and coding of continuous-tone still images – Requirements and guidelines". Http://www.itu.int/rec/T-REC-T.81. Retrieved 2011-01-12.
JPEG2K	"ISO/IEC 15444-1:2004 - Information technology -- JPEG 2000 image coding system: Core coding system". Http://www.iso.org/iso/iso_catalogue/catalogue_ics/catalogue_detail_ics.htm?csnumber=27687. Retrieved 2009-11-01.
WSQ	"WSQ Grayscale Fingerprint Image Compression Specification" Version 3.1. Https://www.fbibiospecs.org/docs/WSQ_Grayscale_Specification_Version_3_1.pdf. Retrieved 2010-01-11.

Appendix A. Compression Ratio

JPEG 2000 implementations, such as OpenJPEG v1.4 used in the present study, holds compression ratio very close to that specified. Thus, for the target bit rate, 0.55 [13] bpp, the compression ratio is 8/0.55 = 14.545:1. Indeed, where this compression ratio is given as input to the OpenJPEG CODEC, the average compression ratio for the dataset of 2898 fingerprint images is 14.690 with a standard deviation of 0.091.

WSQ, by contrast, accepts a bit rate as input, but the measured compression ratio could vary considerably. In the present case, for example, the average compression ratio for WSQ is 17.137 with a standard deviation of 2.425.

Thus, for Case A, in which WSQ and OpenJPEG compression are being compared directly, we first compress using WSQ and measure the effective compression ratio of the output relative to the input image size. Then we apply the OpenJPEG CODEC to the original image using this compression ratio. This procedure ensures that comparable levels of compression are applied by each of the two CODECs. Using this procedure, the normalized difference between compression ratios for each of the N=2898 images is calculated as

$$d_i = \frac{\left| CR_{ii}^{WSQ} - CR^{OPJ} \right|}{\max(CR_{ii}^{WSQ}, CR^{OPJ})}, i \neq 1... \quad (5)$$

Table 6 summarizes the sample median compression ratios and 95 % confidence limits of the estimated medians for both independent compression using the two CODECs at a target bit rate of 0.55 bpp and for constrained compression using the JPEG 2000 CODEC as described above. The median of the difference metric of equation 6 is also shown with confidence limits. Other statistics are also included in Table 6 describing relative distributions of compression ratios and of the normalized difference metric.

Confidence intervals for the medians were estimated using a bootstrap method in which medians are computed for each of 1000 replicates of the data, randomly sampled with replacement, from the original set of 2898 measurements. The upper and lower bounds of the confidence interval are the 0.025^{th} and 0.975^{th} quantile of the resulting distribution of medians.

Applying the two CODECs independently, specifying a target bit rate of 0.55 bpp, yields a median normalized difference (via eq. 6) of 0.098 (9.8 %). Thus, setting compression bit rate independently at 0.55 bpp for the two CODECs could yield actual compression rates differing by more than 10% for the same image.

By contrast, matching the compression rate of the OpenJPEG CODEC to the measured compression ratio measured after application of WSQ to each image yields a median normalized difference between compression ratios of approximately 0.004 (0.4 %).

---

[13] Note that in order to achieve the average compression ratio of 15:1, the FBI typically uses a bit rate specification of 0.75 [BRISLAWN2] rather than the 0.55 used here. We use the published compression ratio of approximately 15:1 (i.e. 0.55 bpp) in order to examine variability of WSQ given this intended compression ratio as a target.

Table 6 - Median compression ratios and % normalized differences with upper and lower 95 % confidence limits (N=2898)

Target Compression Rate set independently to 0.55bpp	Compression Ratio WSQ	Compression Ratio OPJ	% diff WSQ-OPJ
Median (lower limit, 95 % CI)	16.25	14.69	9.34%
Median	16.31	14.69	9.78%
Median (upper limit, 95 % CI)	16.38	14.70	10.24%
Minimum	12.55	14.56	14.53%
Maximum	30.49	15.39	51.99%
Mean	17.14	14.71	12.65%
Standard Deviation	2.43	0.10	11.05%
WSQ compression at 0.55bpp and OPJ rate set to measured CR of WSQ output	Compression Ratio WSQ	Compression Ratio OPJ	% diff WSQ-OPJ
Median (lower limit, 95 % CI)	16.25	16.36	0.41%
Median	16.31	16.42	0.38%
Median (upper limit, 95 % CI)	16.38	16.48	0.36%
Minimum	12.55	14.56	4.82%
Maximum	30.49	30.51	0.03%
Mean	17.14	17.23	0.58%
Standard Deviation	2.43	2.41	0.58%

Appendix B. Effect of Image Contrast on SIVV Power Signal

To demonstrate the effect of contrast on the spectrum (SIVV signal) of an image (see Figure 12, adaptive histogram equalization was applied to a non-compressed fingerprint image to boost the contrast. The two images are compared using the SIVV signal representation The SIVV signal for the contrast enhanced image is raised over the entire frequency range. Uniform increase or decrease in power over all frequencies is more likely due to increase or decrease in overall image contrast than to changes in the image structure. Nill [NILL2] discusses this in the context of the AC and DC components of the image, where a measure of image contrast is given as

$$C_{img} = \frac{\sigma_{img}}{\mu_{img}} \qquad (6)$$

where σ_{img} and μ_{img} are the standard deviation and mean respectively of the image pixel gray levels.

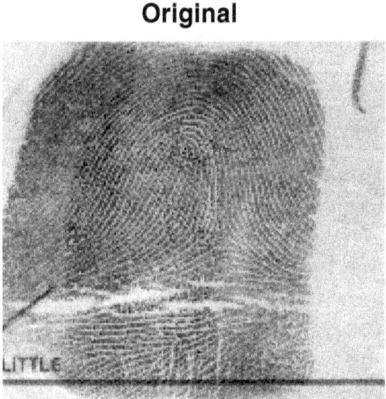 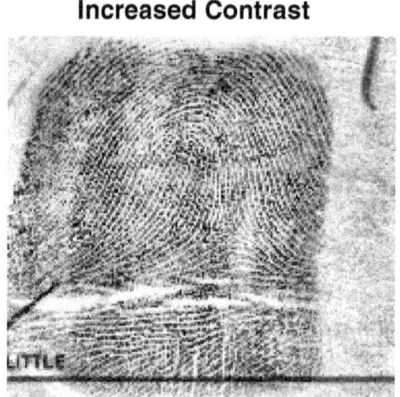

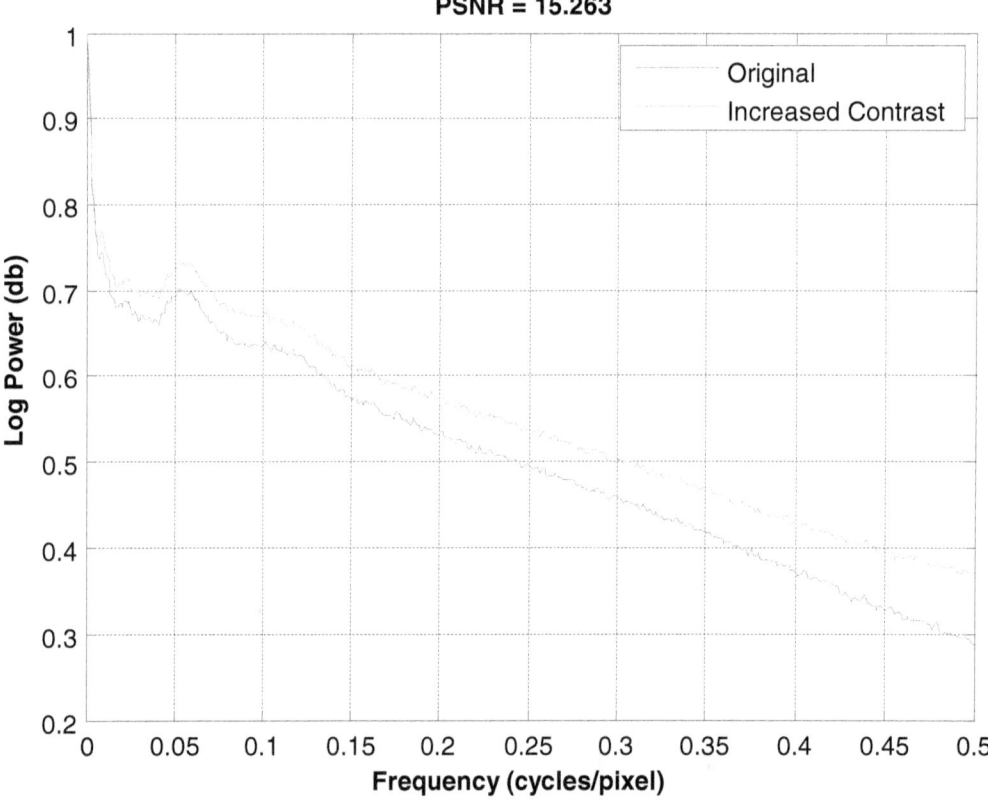

Figure 12 – Non-compressed image (upper left) has had its contrast enhanced using an adaptive histogram equalization algorithm (upper right). The lower display of the SIVV signals shows the almost uniform increase in power over all frequencies. Note the low PSNR value as a result of this operation in spite of the high correlation of the frequency structure of the two images.

Appendix C. Distributions of Comparison Measurements

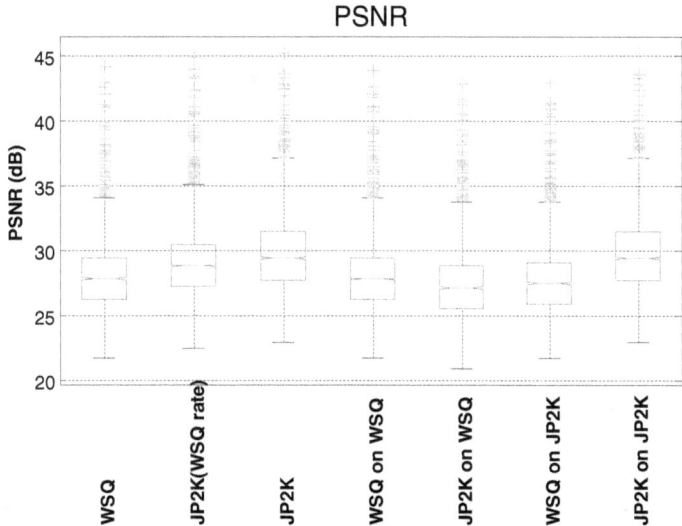

Figure 13 - Boxplots of distributions of PSNR for each compression experiment.

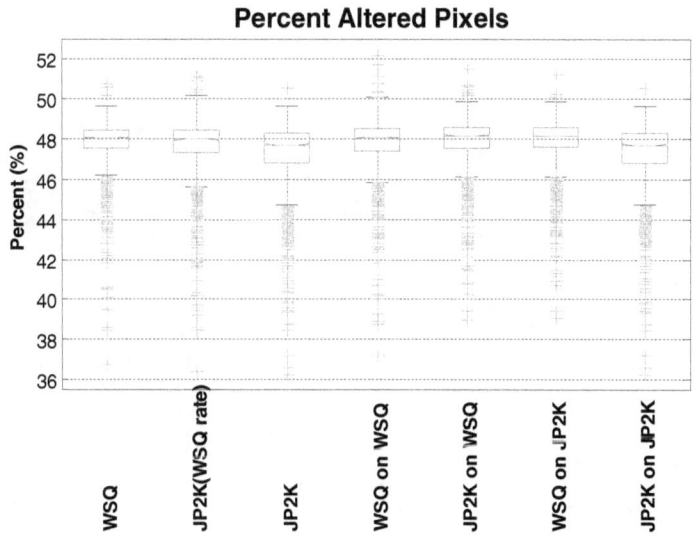

Figure 14 - Boxplots of distributions of proportion of altered pixels for each compression experiment.

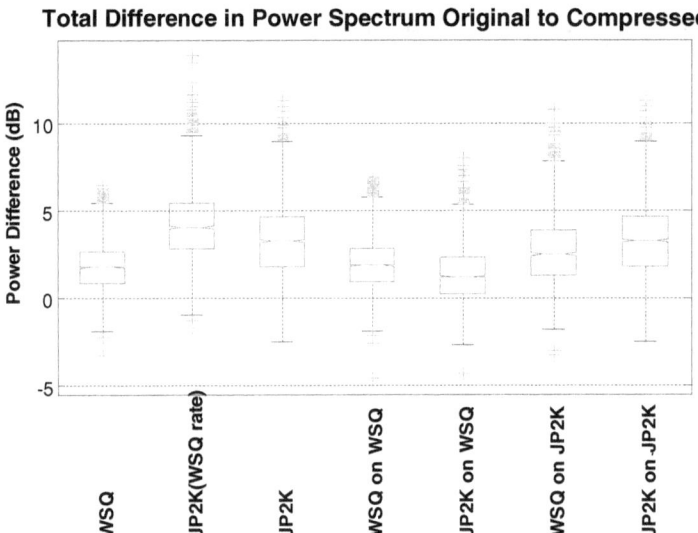

Figure 15 - Boxplots of distributions of total difference in power spectrum for each compression experiment.

www.ingramcontent.com/pod-product-compliance
Lightning Source LLC
Chambersburg PA
CBHW081740170526
45167CB00009B/3890